PRAISE FOR *THE PERFECT SWARM*

"This would be my nominee for book of the year, if it wasn't still only January. Who knows what may turn up in the next 12 months? Whatever it is, though, will find Fisher a hard act to follow." —*BBC Focus* Magazine

"In *The Perfect Swarm*, science popularizer Len Fisher ranges far and wide to discover what tips the science of complexity has for us." —*New Scientist*

"From making a quick exit to detecting fraud statistically, Fisher has you covered." —*Post & Courier*

"Len Fisher reveals how study of animal swarms allows us to better understand our own society. By blending personal stories with a clear presentation of new theoretical ideas he shows why rumors, ideas and information spread so rapidly through groups."

—David Sumpter, professor of mathematics, Uppsala University

"That complexity can be simple to explain might seem counterintuitive, but in *The Perfect Swarm: The Science of Complexity in Everyday Life*, Len Fisher demonstrates just that. This book provides a thoughtful, entertaining, and—most important—easy to understand treatment of how patterns emerge and problems can be solved when many individuals interact in very simple ways. Clear and fluent, *The Perfect Swarm* is an enjoyable source of insight for those who would like to better understand how many seemingly complex things in the world really aren't so complex after all."

—Gregory Sword, associate professor of biology, University of Sydney

"*The Perfect Swarm* does a marvelous job of explaining the network effects that determine our daily life. I highly recommend it to anybody seeking practical solutions to the puzzling complexities of everyday life, and especially to anyone interested in the mathematical and physical underpinnings of swarm intelligence, swarm business, and swarm creativity."

—Peter A. Gloor, author of *Swarm Creativity* and *Coolhunting*

"From locusts watching *Star Wars* to Murphy's Law of Management, *The Perfect Swarm* hits all the buttons. This is a wonderful tour through the new mathematics of swarms, flocks, and crowds, and it makes the emerging science of complex systems seem simple. Easy to read and highly informative."

—Ian Stewart, author of *Professor Stewart's Cabinet of Mathematical Curiosities*

"I am not sure there is a 'science of complexity,' but there undoubtedly are a lot of interesting ideas emerging about underlying simplicities, and their implications, within many seemingly complicated systems. Len Fisher's book is a truly excellent and clearly-written guide to this exciting area."

—Lord Robert M. May, Zoology Department, Oxford University

"*The Perfect Swarm* is the perfect introduction to the intelligence of groups and the power of simple decision rules."

—Gerd Gigerenzer, Director, Max Planck Institute for Human Development, author of *Gut Feelings*

THE PERFECT SWARM

ALSO BY LEN FISHER

Rock, Paper, Scissors

Weighing the Soul

How to Dunk a Doughnut

THE PERFECT SWARM

The Science of Complexity in Everyday Life

Len Fisher, Ph.D.

BASIC BOOKS
A Member of the Perseus Books Group
New York

Hardcover first published in 2009 by Basic Books,
A Member of the Perseus Books Group
Paperback first published in 2011 by Basic Books

Designed by Timm Bryson

The Library of Congress has catalogued the hardcover as follows:
Fisher, Len.
The perfect swarm : the science of complexity in everyday life / Len Fisher.
p. cm.
Includes bibliographical references and index.
ISBN 978-0-465-01884-0 (alk. paper)
1. Swarm intelligence—Social aspects. 2. Group decision making. 3. Group problem solving. 4. Social groups—Psychological aspects. I. Title.
HM746.F57 2009
302.3—dc22
2009031018

Paperback ISBN: 978-0-465-02024-9

10 9 8 7 6 5 4 3 2 1

To Wendella, who has now survived four books,
and has helped me to do likewise.

CONTENTS

PATTERNS OF THE PERFECT SWARM: VISIONS OF COMPLEXITY IN NATURE

How Complex Patterns Emerge from Simple Rules in Physical and Living Systems

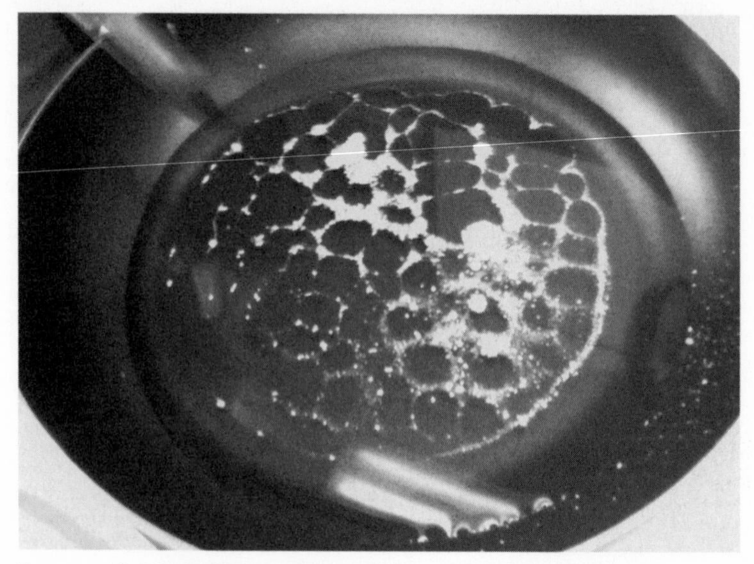

Pattern of Rayleigh-Bénard cells formed by convection in a layer of oil in a frying pan heated from below (see page 3).

COURTESY OF BEN SCHULTZ

Patterns in coral at Madang, New Guinea, formed by walls of calcium carbonate secreted by individual polyps competing for space.

PHOTO BY JAN MESSERSMITH

Pattern of reaction products formed in a petri dish by an "oscillating" chemical reaction known as the Belousov-Zhabotinsky reaction. The black dots are adventitious air bubbles.

COURTESY OF ANTONY HALL, WWW.ANTONYHALL.NET

Patterns formed by tens of thousands of the soil-dwelling "slime mold" amoeba *Dictyostelium discoideum*, growing on the surface of a gel-filled petri dish. Each individual is responding to chemical signals from its neighbors that warn of a lack of bacteria that are its main food. Ultimately the cells will aggregate to form a "slug," technically called a grex, which can crawl through the soil much faster than the individual amoebae to find new bacterial pastures (see page 18).

COURTESY OF PROF. CORNELIS WEIJER, UNIVERSITY OF DUNDEE, UK

Stripes in the Algodones sand dunes of Southern California, formed by a combination of wind driving the sand up and the force of gravity pulling sand grains down (see page 2).

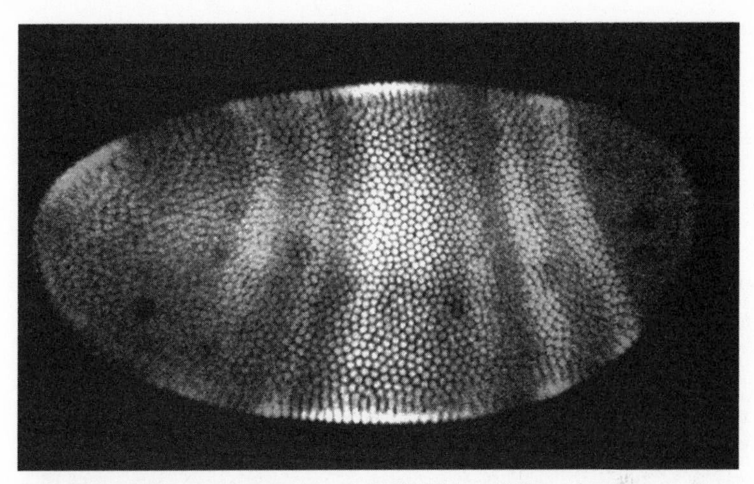

Stripes in the developing larva of a fruit fly (*Drosophila melanogaster*). The stripes are formed by the selective differentiation of cells in response to the presence of distinct neighbors. Each stripe will ultimately develop into a different part of the adult body—wings, thorax, eyes, mouth, etc.

COURTESY OF JIM LANGELAND, STEVE PADDOCK, SEAN CARROLL. HHMI, UNIVERSITY OF WISCONSIN

Stripes on a tiger, once thought by some biologists to emerge from a balance of chemical gradients in a manner reminiscent of the Belousov-Zhabotinsky reaction. This tiger was photographed in Pench National Park, Madhya Pradesh, India in 2004.

Spiral pattern in a sunflower head, an arrangement that allows for "optimal packing" of the individual parts. This design is described mathematically by the Fibonacci sequence, in which each number is the sum of the two previous numbers (i.e. 0, 1, 1, 2, 3, 5, 8, 13, and so on).

COURTESY OF DR. HOWARD F. SCHWARTZ, COLORADO STATE UNIVERSITY, WWW.BUGWOOD.ORG

Spiral galaxy, whose formation is dominated by Newton's Law of Gravity and his three laws of motion. Barred Spiral Galaxy NGC 1300; image taken by Hubble telescope.

COURTESY OF NATIONAL AIR AND SPACE ADMINISTRATION (HTTP://HUBBLESITE.ORG/GALLERY/ALBUM/ENTIRE/PR2005001A/NPP/ALL/)

Self-organization in a school of fish, produced by each animal following Reynods' three laws. Shoal of Robust Fusilier, Caesio cuning, German Channel, Micronesia, Palau (see page 12).

Self-organization in a herd of wildebeest crossing the Serengeti plain.

Self-organization produced by "social forces" of people spontaneously forming "lanes" as they walk along a crowded street (see page 50).

COURTESY OF DIRK HELBING

ACKNOWLEDGMENTS

It is a pleasure to acknowledge the many people who have acted as advisors, mentors, and muses in my efforts to produce a simple guide to complexity. Amanda Moon, my editor at Basic Books, has done her usual extremely thorough and helpful job, as has Ann Delgehausen, who has performed marvels with the copyediting. Particular thanks must also go to my wife, Wendella, who has painstakingly examined every chapter from the point of view of the nonscientific reader, suggesting a multitude of interesting examples and pointing out where new ideas needed a clearer explanation.

The book has also benefited greatly from the advice of world experts in the various fields that it covers. I have named them individually in the notes to the appropriate chapters and here record my collective thanks, along with my thanks to other friends and colleagues (scientific and otherwise) who have gone to a great deal of trouble to read the manuscript and offer suggestions that have contributed considerably to its gradual improvement over the course of writing. I can't blame any of them for errors that may still have crept in. Those, unfortunately for my ego, are my responsibility alone.

Here are the names of the people who helped, in alphabetical order, accompanied by the offer of a drink when we next meet: Hugh Bray, Matt Deacon, John Earp, David Fisher, Gerd Gigerenzer, Dirk

Helbing, Jens Krause, Michael Mauboussin, Hugh Mellor, James Murray, Sue Nancholas, Mark Nigrini, Jeff Odell, Harry Rothman, Alistair Sharp, David Sumpter, Greg Sword, and Duncan Watts.

If I have omitted anyone, I can only apologize, and extend the offer to two drinks.

Introduction

Shortly after *Star Wars* hit box office records, a group of ninety-seven locusts sat down to watch the film. They didn't have much choice in the matter; they were strapped in place with their heads firmly clamped while experimenters monitored spikes of electrical activity in their brains as they reacted to the fleets of spaceships zooming at them from either side.

The scientists were trying to work out how locusts in a dense swarm manage to avoid colliding with each other. Studies on this aspect of swarm behavior have provided valuable information about our behavior in the human swarm, from working our way through crowds to the design of collision avoidance systems for cars. There are many other lessons that we can also learn from the behavior of animals in groups, such as swarms of locusts, flocks of birds, and schools of fish. This book is about how we can use such lessons to make better group decisions and better decisions for ourselves as individuals within a group.

The individual animals in a swarm, flock, or school follow rules that help them to get the most from the group. Some of these rules help them to stay together as a unit. Others allow them to act as if they were components of a superorganism, which has no individual leader, and where the whole becomes greater than the sum of its parts as the group develops swarm intelligence and uses it to make collective decisions.

The modern science of complexity has shown that collective behavior in animal groups (especially those of insects such as locusts, bees, and ants) emerges from a set of very simple rules for interaction between neighbors. It has also revealed that many of the complex patterns in human society arise from similarly simple rules of social interaction between individuals. My ultimate aim in this book is to explore how the process works and, more importantly, to help find simple rules that might guide us through the fog of complexity that so often seems to enshroud our lives.

The process by which simple rules produce complex patterns is called "self-organization." In nature it happens when atoms and molecules get together spontaneously to form crystals and when crystals combine to form the intricate patterns of seashells. It happens when wind blows across the sands of the desert to produce the elaborate shapes of dunes. It happens in our own physical development when individual cells get together to form structures such as a heart and a liver, and patterns such as a face. It also happens when we get together to form the complex social patterns of families, cities, and societies.

There is no need for a central director to oversee the process. All that is needed is an appropriate set of simple local rules. Individual sand grains form themselves into dunes under the combined forces of gravity, wind, and friction with nearby grains. Atoms and molecules experience forces of attraction and repulsion with nearby atoms and molecules, and these forces are sufficient in themselves to produce long-range order that can extend for billions of atomic diameters in all directions.

Our society is made up of billions of individuals also, and the forces of attraction and repulsion between us can act to create social structures among us as well. These structures, however, are not nearly as regular as those of atoms in a crystal. To use the complexity scientist's picturesque but slightly misleading phrase, they are on the *edge of chaos.*

The meaning of the phrase can be unclear because *edge* implies that our social structures are forever in danger of descending into anarchy. What it really means is that their *degree* of organization lies somewhere between complete order and complete chaos.

Complete chaos is rather hard to achieve, although my wife claims that the disordered piles of paper on my desk come pretty close. I argue that there is order within the chaos, even if I am the only one who can see it.

There is order within most forms of chaos. This is *dynamic* order, which you can see by simply pouring some cold milk into a cup of hot black coffee. Patterns appear on the surface that are an indication of what is going on underneath, where the mixing of the hot and cold liquids produces a set of swirling vortices that rapidly self-organize into a remarkably regular arrangement. These are called Rayleigh-Bénard cells. You can find them in a fraction of an inch-thick layer of liquid in a shallow dish and in the miles-thick layer of the Earth's atmosphere.

Systems on the edge of chaos, including animal groups and human societies, also have dynamic order, but it lasts a lot longer than the vortices in a cup of coffee. The order arises from rules of interaction between individuals that produce large-scale dynamic patterns of interaction. The resulting set of emergent patterns characterizes the society as a whole rather than its individual members.

These patterns span a variety of time scales. Some, such as those of cities, can be very long lasting. Others, such as those of a moving crowd, may be as evanescent as clouds in a windy sky. Still others, such as those of human relationships, can be anywhere in between.

Two sorts of dynamic pattern are possible in a system on the edge of chaos. In one, the system cycles endlessly between different positions, as sometimes happens in domestic arguments that go round and round without any resolution. In the other pattern, a much more productive one, the system adapts to meet changing circumstances, as does the shape of a school of fish when confronted by a predator.

When individuals in a group are able to respond collectively to changes in circumstances, the group becomes a complex adaptive system. The rules that produce such systems and govern their behavior are of considerable interest, and not just to students of human society but also to students of the whole of nature.

Successful ecosystems are complex adaptive systems, as are successful cities and societies. According to the scientist James Lovelock's Gaia concept, the Earth as a whole is a complex adaptive system. One of its long-term adaptations that should be of concern to all of us may well be to get rid of our species to protect itself. Whether that happens or not could come down to whether we are able to understand the rules that govern its complexity, and whether we have the wisdom to adapt ourselves and conform to those rules.

For a complex adaptive system to evolve and grow, the interactions between its individual members must be of a special kind. Instead of being simply additive, in the manner of a number of individuals pulling on a rope in a tug-of-war competition, the interactions must be nonlinear, meaning that an action by one individual produces a disproportionate response in other individuals or in the group as a whole. Just one person clapping at the end of a concert, for example, can stimulate several others to start clapping, and they in their turn can each stimulate a number of others until soon the whole audience is applauding.

Collective clapping can sometimes fall into synchrony, which is a property of the audience as a whole, not of any individual member. Such emergent properties arise in complex adaptive systems when there are several types of nonlinear action going on at once.

One of the most important emergent properties a group can have is swarm intelligence, which allows a group to tackle and solve problems in a way that its individual members cannot. In this book I examine the simple rules that give rise to swarm intelligence in nature. I ask whether we can use swarm intelligence and its underlying rules (or other equally simple ones) to help us to steer our way through the complexities of life.

Our journey of discovery begins with the animal kingdom and progresses in nine stages. The first three cover the evolution of swarm intelligence in the natural world and what we can learn from its underlying rules. The following four focus on developing group intelligence in human society and using it to solve complex problems. The final two look deep into complexity itself to uncover new and simple rules that we can use to make the best decisions we can when we are trapped in its web.

Chapter 1 is an overview of swarm intelligence: What is it? How does it arise from nonlinear interactions? What sorts of animals use it? What advantages does it convey to the individuals within a group and to the group as a whole?

The following two chapters cover the rules that locusts and bees use when flying in swarms, and the land-based logic of ants. These three types of insect use the basic rules of complex adaptive systems to implement swarm intelligence in very different ways. We can learn something from each of them.

The collision avoidance strategies of locusts have implications for driving in traffic and walking in crowds. Bees use "invisible leaders" to direct the movements of the swarm. We can do the same, and also take advantage of such leaders when traveling in unfamiliar surroundings. Ants use a specialized form of group logic that allows them to find shortcuts and optimum routes. We can follow their example while walking or driving. You may be surprised by the ways their problem-solving approach is being applied in many other situations.

After the chapters on insect logic, I look at individual behavior in human crowds and describe how recent research into complex crowd dynamics has revealed optimum strategies for making our way through them and handling ourselves in dangerous crowd situations.

In the next two chapters I focus on group decision making. In the first I ask whether we should follow an average course, one that takes equal account of everyone's opinion, or whether we should go with one endorsed by the majority. In the second I show how we can use group intelligence to achieve the best consensus and how we can best avoid the perils of groupthink while doing so.

One way to implement group intelligence is through networking. In chapter 7 I explore different methods of networking, including those that lead to the famous six degrees of separation. I show how new understanding has led to more effective strategies for networking and communication, and contributed to our ability to prevent the spread of disease through the human network.

The penultimate chapter is devoted to the ways in which we can use simple rules to make the best individual decisions when we are confronted by complex problems. Some of the best approaches are very simple indeed and provide surprising insights into the amount and type of information that we need to make the best decisions.

Finally, in chapter 9 I explore one further way in which we might make decisions: by looking for patterns within the complexity. Sometimes these can guide us in the right direction, but, as the science shows, the whole of society is frequently greater than the sum of the parts and we need to be aware of times when overall complexity can overwhelm the simplicity that lies buried within. Simplicity is OK, but complexity rules. OK?

A Note on the Notes I uncovered many fascinating anecdotes, references, and points of interest during my research for this book that did not quite fit into the main text without disrupting the flow of the story. I have put these into a set of notes that are designed to be dipped into, enjoyed, and read quite independently of the main text. Several readers of my previous books, which I arranged the same way, have written to me to say that the notes section is where they start! If you would like to know more about any of the topics mentioned in the main text, there is very likely more detailed information in the notes.

Some of the references point you to the underlying scientific literature (usually accessible through websites such as Google Scholar). I have done my best to choose articles that are both seminal and easily readable.

Enjoy!

ONE

The Emergence of Swarm Intelligence

The behavior of animals in a swarm used to be seen as almost magical. Some early scientists even thought that swarms of insects, schools of fish, and flocks of birds could produce their wonderfully coordinated movements only through some sort of extrasensory perception, or perhaps through the development of a group consciousness that required each animal to sacrifice its individuality and become a puppet of that consciousness.

Animal behaviorists, informed by the science of complexity, have now proved that swarm behavior does not need such *outré* explanations. It emerges naturally from simple rules of interaction between neighboring members of a group, as happens with a wave generated by a crowd of spectators at a football game. The wave might look to a visiting Martian like a complicated exercise in logistics, but its dynamic pattern emerges from a simple rule: stand up and stick your hands in the air (and then put them down again) as soon as you see your neighbor doing it.

Such a wave involves rapid transmission of information from individual to individual, and this is a key feature of swarm behavior. It happens in the human swarm in the form of gossip—neighbor chats with neighbor and additional information is channeled back along the same route until everyone knows what is going on and can act on the information. My wife and I once turned up at a country fair, to which we had been invited by friends, and were greeted at the gate by a perfect stranger who took one look at us and said, "Your friends are over in the beer-tasting tent." She hadn't actually seen our friends, but she had heard through the grapevine where they were and that they were expecting people who fit our description.

Swarm behavior becomes swarm intelligence when a group can use it to solve a problem collectively, in a way that the individuals within the group cannot. Bees use it to discover new nest sites. Ants use it to find the shortest route to a food source. It also plays a key role, if often an unsuspected one, in many aspects of our own society, from the workings of the Internet to the functioning of our cities.

Swarm intelligence is now being used by some people in surprising and innovative ways. Companies are being set up that are run by swarm intelligence. Computer programmers are using it in a radical approach to problem solving. There is even an annual event, Swarmfest, where scientists swarm together to discuss new applications of swarm intelligence.

Groups that use swarm intelligence need no leader, and they have no central planning. What, then, allows them to maintain their coherence and to make seemingly rational decisions? How do individual interactions translate into such complex patterns of behavior? To make the best of our own individual interactions, we need to understand the answers to these questions. The answers have come from three sources: the real world of animals, the imaginative world of science, and the virtual world of the computer. Here is a brief background for each one.

Learning from the Animal Kingdom

Animals use swarm intelligence to hunt for food and find shelter as a group, and to avoid predators. The scientific study of their behavior, ethology, has uncovered the simple rules they use to engender swarm intelligence. It has also caused the scientists concerned to face some unusual perils on occasion.

German ethologist Martin Lindauer was caught in a particularly bizarre situation in the mid-1950s while he was trying to understand how honeybee swarms find their way to new nest sites. His practice had been to run along underneath a swarm through the outer suburbs of war-ravaged Munich, always wearing a white lab coat while doing so, perhaps to publicize his scientific credentials. Unfortunately, his coat resembled the uniform that patients in a nearby hospital for the dangerously insane were forced to wear. One day, guards from the hospital mistook him for an escaped patient and gave chase. Luckily, he was able to run faster than the guards, which showed not only how fit he was but also how fast swarms of bees can fly.

We owe much to Lindauer, and to other ethologists who have exposed themselves to risk in the cause of science. When two Brazilian scientists decided to follow schools of piranhas by snorkeling directly above them, they dismissed the dangers of being attacked by the fish and were probably right to do so. They were on less sure ground in casually dismissing the danger of attack from caimans that were hunting nearby. With typical scientific understatement, they simply complained in their report that the caimans hampered their nighttime observations by muddying the waters with the lashing of their tails.

The Brazilians were not the first to use snorkeling as a means of following schools of fish. That honor appears to belong to the Greek philosopher Aristotle, who is believed by some historians to have donned a face mask and thrust his bearded visage under the waters of the Aegean Sea to observe that "sea basse [*Dicentrarchus labrax*] and the grey mullet school together despite the hostility between the kinds."

But Aristotle risked no more than a wet beard. A scientist with whom I was studying coral reef ecology on Australia's Great Barrier Reef risked rather more when he teased a supposedly harmless gummy shark that was lying on the bottom by poking it with his flipper. He explained to us that the shark had weak jaws and tiny blunt teeth. The shark proved him wrong on both counts by biting though his flipper and hanging on to it grimly. The water was 6 feet deep; the scientist was 5'10". The only way that he could escape drowning was to bend down, undo his flipper, and leave it with the shark.

All of these scientists made original discoveries about the animals they were studying. University of Miami biologist Brian Partridge, however, was the first scientist to make real progress in understanding how groups of animals can sometimes move, act, and make decisions as though they were a single superanimal. His chosen species was the saithe.

Saithe are also known as pollock (or pollack), and they are increasingly finding their way onto Western menus following the decline of the cod and haddock fisheries. There are two species: Alaskan pollock (*Theragra chalcogrammais*), said to be "the largest remaining source of palatable fish in the world," and Atlantic Pollock (*Pollachius virens*). Both are around three feet long and weigh up to forty-five pounds.

Partridge was studying Atlantic pollock. Like many fish, it exhibits schooling behavior. Partridge realized that to understand how the school stayed together and moved as a unit he would have to be able to identify and follow every individual fish.

The first bit was easy—he simply branded each fish on the back (using a freezing brand rather than a red-hot one to make the mark). Following the fish was going to be rather more difficult. To do it he arranged to get schools of twenty to thirty saithe swimming around a thirty-three-foot diameter doughnut-shaped tank at the University of Aberdeen in Scotland. While the fish were swimming, the experimenter lay on a rotating gantry above the tank and followed the movement of the school, recording a continuous race-track commentary on how the individual saithe were performing. Since the school was swimming at around one foot per second, this meant that the experimenter was being swung around, head down, at one revolution per minute. This doesn't sound like much, but when I reproduced it on a merry-go-round at a child's playground, I found it to be a fairly dizzying experience.

Dizziness, however, was the least of the experimenter's worries. After the experiments had been completed and the fish released (or eaten—his paper does not say which), Partridge and his helpers sat down to painstakingly measure the relative fish positions in more than twelve thousand frames of film. At the end, they discovered the key rules that permitted the school to move as a unit. There were just two, which each fish obeyed: follow the fish in front (if there is one*) and keep pace with the fish beside you.

These two simple rules, expressed in various forms, are now known to underlie all sorts of complex group movements, from the wonderfully unified flashes caused by changes of shape or direction in schools of fish to the movements of flocks of birds, swarms of insects, and crowds of humans. But how does the complexity actually arise? What processes are involved? To find answers to these questions we must turn to another source, the world of physical science.

Learning from Science

My first encounter with the application of science to the problems of complexity came while I was playing bridge in the early 1970s.

---

* If there isn't one, then obviously only the second rule applies!

My partner was Robert (now Lord) May, and I couldn't understand the squiggles and marks he was making in a notebook on the corner of the table when he was not playing a hand. I had no idea at the time that he was making history.

Robert was puzzled by the behavior of a very simple equation called the "logistic difference equation," which mathematicians use to describe how animal populations grow. It was a perfectly respectable equation, and it gave perfectly respectable answers. It predicted, for example, that populations would initially grow exponentially but that when food, space, or other resources became limited the population would plateau at a level that the environment could sustain.

Robert had discovered a paradox, though. At certain rates of population growth, the equation went crazy. Instead of predicting smooth changes, it predicted cyclic or chaotic transitions between boom and bust, indicating that populations could appear to thrive but then suddenly crash. The equation produced these results because it contained elements of positive and negative feedback, elements that are now known to be central to the emergence of all sorts of complexity, including dramatic population fluctuations in nature, equally dramatic fluctuations in the stock market, and stable patterns such as those involved in swarm intelligence.

Positive feedback is a cyclic process that is responsible for the squealing of a PA system at a concert when the amplification is turned up too high: the sound from the speakers is picked up by the microphone, which feeds it back to the speakers through an amplifier that makes it even louder, which sends it back to the speakers in a vicious circle that eventually overloads the system so that it produces a howl of protest. The scientific intelligence expert R. V. Jones observed a wonderful example at a lonely airfield during World War II. A microphone and loudspeaker had been set up on opposite sides of the landing strip, and someone happened to laugh near the microphone. The amplifying system was just on the edge of positive feedback, and the laugh was very slowly amplified after the person had walked away, leading Jones to speculate that a human was no longer needed, and here was a machine that could laugh by itself.

The credit crisis that began in 2008 provides a less humorous example of the effects of positive feedback, which in this case amplified a mistrust of financial institutions until the worldwide financial system was in danger of collapse. Many individual financial institutions have already failed under the stresses caused by positive feedback, which makes its presence known in the form of a run on these institutions. One example is the collapse of Washington Mutual on September 25, 2008. Over ten days, more and more investors had realized that others were withdrawing their money, and they rushed to withdraw theirs as well. The total reached $16.7 billion.

Strong preferences, or fashions, can also arise from seemingly insignificant beginnings through the operation of positive feedback on a small random fluctuation. Say, for example, that most of your friends own either a Ford or a Toyota, and you are trying to choose between the two when it comes time to buy a new car. You ask around, and just by accident the first three people you encounter own a Ford and are perfectly happy with it. So you buy a Ford.

After you buy a Ford there is one more person in the group who owns a Ford, which slightly increases the chance that the next person who asks around will talk to Ford owners rather than Toyota owners. If she buys a Ford, too, there will be two more people in the group who own Fords. The "Ford effect" amplifies, and eventually most of the group owns Fords. (It would have been the "Toyota effect" if you had talked initially to several people who owned Toyotas.)

The pattern of everyone owning a Ford (or a Toyota) is one that has emerged from the application of a simple local rule (choose the car that the first three people that you meet own and are happy with) together with the action of positive feedback on a chance fluctuation (the first three happened to own the same brand).

Positive feedback is not the only way to produce a runaway effect. Such effects can also arise from a chain reaction, such as the one described by James Thurber in "The Day the Dam Broke," part of his autobiography. Triggered initially by the sight of just one person running, the entire citizenry of the East side of Columbus, Ohio, fled from a nonexistent tidal wave, despite reassurances that there was no cause for alarm. Thurber and his family were among those running. "We were passed," says Thurber, "in the first half-mile, by practically everyone in the city." One panicking citizen even heard the sound of "rushing water" coming up behind him; it turned out to be the sound of roller skates.

The panic arose because the sight of the first person running got several other people running, and the sight of each of these got a few more people running, and so on. The process continued until everyone was running. It is the same process that is at work in an atom bomb, when the disintegration of an atomic nucleus releases energetic neutrons with enough energy to break up several nearby nuclei, and each of these produces enough neutrons as it disintegrates to break up several more. The ongoing cascade produces rapid exponential growth in the number of neutrons and the amount of energy being released until there is a giant explosion.

The chain reaction is controlled in nuclear power stations by the insertion of cadmium rods into the disintegrating material. The rods absorb a sufficient number of neutrons to block the chain reaction and produce a controlled release of energy instead. One of the great discoveries of complexity science is that a similar stabilizing outcome can be produced in many social situations by introducing negative feedback to counteract the destabilizing effects of chain reactions and positive feedback.* The result is a complex dynamic pattern with its own inherent stability, but also with the potential for evolution and growth.

---

* To avoid confusion, we should note that the physical scientist's meaning for the terms positive and negative feedback are not the same as those of a psychologist. To a psychologist, negative feedback is destructive and destabilizing, while positive feedback is good and desirable. To a physical scientist, the implications are usually the reverse.

Negative feedback exerts its balancing effect by acting to preserve the status quo. A simple example is a governor on a motor, which acts to progressively reduce the rate at which fuel is supplied as the engine speeds up, so that the engine can never run out of control.

Negative feedback is frequently used to "correct" errors. When an error starts to creep in, the change from the status quo initiates a feedback process that acts to correct the error. When you are driving your car, for example, and you start to drift slightly to the right, your brain automatically applies negative feedback to get you to turn the steering wheel slightly to the left so as to bring you back on course. Positive feedback, which progressively amplifies small effects, would have you turning the wheel farther to the right and sending you farther off course.

In economics, Adam Smith's concept of the invisible hand, which says that the marketplace is self-regulating and will always return to equilibrium after a disturbance, is based on the idea that the institution of the marketplace has built-in negative feedback. As we shall see, modern complexity theory recognizes that this is far from being the case in practice, and that our complex economic system is governed by an intricate balance of positive and negative feedback, with the occasional chain reaction thrown in.

The balance ultimately depends on the rules of interaction between individuals (these rules are known technically as "behavioral algorithms"). Two of the key problems in understanding the emergence of collective properties like swarm intelligence are identifying the patterns of interaction that individual animals (including ourselves) follow and detecting how information flows between the animals. Much of this book is concerned with the former, and working out how we can use these patterns to our advantage.

For a group to have collective adaptability (the ability of the group as a whole to respond to changing circumstances) nonlinear rules alone are not usually sufficient. Complexity theorists John Miller and Scott Page list a total of eight criteria for collective adaptability, based loosely but respectfully on the Buddhist eight-fold path:

Right View The individuals in a group (complexity scientists call them "agents") must be able to receive and make sense of information from other individuals in the group and from the world in general.

Right Intention Agents must have some sort of goal that they want to achieve. Fish may want to avoid being eaten, for example, while people may want to act collectively to achieve political change.

Right Speech Agents must be able to transmit information as well as to receive it. This need not involve actual speech. Cells in the communal slime mold *Dictyostelium discoideum*, for example, communicate by sending chemical messages, and the neurons in our brains communicate via electrical impulses.

Right Action Agents must be able to influence the actions of nearby agents in some way.

Right Livelihood Agents must receive some sort of payoff for their actions within a group, such as a salary for a task performed or the threat of a punishment like dismissal if the task is not performed.

Right Effort Agents need strategies that they can use as they anticipate and respond to the actions of others.

Right Mindfulness There are many kinds and levels of rationality. Our task as agents in a complex society is to choose and use the right level of each.

Right Concentration To understand how complexity emerges, we sometimes have to go back to the old scientific approach of concentrating on one or two important processes, temporarily ignoring the rest.

All of these criteria are covered in the pages that follow, sometimes in very different contexts. Right mindfulness, for example, covers the level of detail that we need to have to make good individual decisions and also the ways of thinking that we need to adopt to reach consensus as a group.

Virtual Worlds

To understand how these criteria influence our choice of behavior in complex situations, we often need to resort to computer modeling. Predicting their consequences can be virtually impossible without the aid of such simulation, both for practical and ethical reasons.

One practical reason is that the human mind simply cannot encompass all of the variations and variability inherent in complex adaptive systems. This is why science has progressed almost exclusively in the past by making severe simplifications that allow us to abstract the essentials of a problem. Even when it comes to the relative movements of the sun, the Earth, and the moon, we can calculate the orbits of any pair around each other only by ignoring the effects of the third body. An exact calculation of the three together (known as the "three-body problem") is beyond our analytical powers, and we have to rely on computer simulations just to get a close approximation.

Interactions in society are more complex, and it is only with the advent of powerful computers that we have been able to model how complexity can emerge from simplicity. Such models are now used to understand aspects of crowd behavior, networking, and other aspects of our complex society. (Studies of crowd behavior in particular frequently preclude experimentation because of ethical concerns, especially if an experiment would involve putting individuals in dangerous situations.)

The models are rather similar to those of games like *Tomb Raider*, where virtual individuals are given specific rules of behavior. In the world of complexity science, though, there is no outside player to control what then happens. Instead, the virtual individuals are released in their virtual world, armed only with rules for interaction, while the programmer watches to see what happens.

The rules might be guesses about how people interact with each other in a crowd, for example, and the outcome would be the behavior of the crowd when the individuals follow those rules. By

The Logistic Difference Equation

The equation below looks incredibly simple at first sight, but it has probably driven more mathematicians crazy than any other equation in history.

It was first applied to population growth. If a population of p individuals can grow without limit, and it does so at a constant rate r, then we can simply write:

$$p_{present} = r \times p_{previous}$$

If the population were growing at 3 percent per year, for example, and the population were measured on the same date each year, then the value of r would be 1.03.

This is called exponential growth, and it is quite clear that our planet cannot support it indefinitely. No matter what adaptations we make, there must be some upper limit. Let's call K the largest population that the Earth could sustain, and follow the clever idea of the Belgian mathematician Pierre François Verhulst, who in 1838 proposed a simple equation to describe the way in which population growth must slow down as it approaches its upper limit, and even become negative if it overshoots that limit.

Verhulst's equation, the logistic difference equation, is:

$$p_{present} = r \times p_{previous}((K - p_{previous})/K)$$

This simple-looking equation (note that it is nonlinear, because $p_{previous}$ gets multiplied by itself) has produced some truly extraordinary insights. If you really tweaking the rules, the programmers can come up with reasonable suggestions for the most productive way for individuals to behave in crowds, and also for the best designs for the environments in which crowds might gather, such as city streets, stadiums, and nightclubs.

One other use of computer programming is to mimic the way in which social animals (particularly insects) use swarm intelligence to solve problems. A swarm of virtual individuals is let loose in the

The Logistic Difference Equation (continued)

hate algebra, just skip the next two paragraphs, but do have a look at what follows.

At first sight, the equation looks really neat. When populations are far from their limit, $p_{previous}$ is much smaller than K, and the equation simplifies to the exponential growth equation. When populations get close to their limit, the growth slows right down as $(K-p_{previous})$ gets closer and closer to zero.

This equation neatly describes the growth of bacteria in a petri dish and algae on a pond (so long as the food or light doesn't run out). If you draw a graph of population total as a function of time, it comes out as a classical sigmoid shape, with exponential growth at the beginning and an asymptotic plateau after a longer time—so long as the rate of growth is not too high.

Everything stays normal until we reach a tripling rate of growth ($r = 3$), and then strange things start to happen. The smooth population growth curve breaks into oscillations between two values that correspond to "boom" and "bust." By the time that the growth rate reaches 3.4495, the curve is oscillating between four values. When the growth rate reaches 3.596, there are sixteen states, with the population oscillating rapidly between them. A little above that, complete chaos ensues.

The mathematics of boom and bust accurately describes many events that happen in the real world. Unfortunately, this doesn't always make them easier to predict (as demonstrated by the history of the 2008 credit crisis), partly because the model is always a simplified system compared to reality, but also because the behavior of the system can depend very sensitively on the precise conditions.

artificial computer environment, but this time the environment is designed to reflect the problem to be solved. The individuals might, for example, be given the task of finding the quickest routes through a network that mimics the arrangement of city streets or the routes in a telecommunications network. Amazingly, the solutions that the swarm comes up with are often better than those produced by the most advanced mathematics.

All of these uses of computer programming, scientific rules, and lessons from the animal kingdom are covered in the following pages. We begin with the lessons that locusts, bees, and ants have to offer. Each of them uses subtly different forms of swarm intelligence, and each of their approaches has something different to tell us about group problem solving in our own world.

TWO

The Locusts and the Bees

Locusts and Us

Locusts are distinguished from other types of grasshoppers because their behavior changes radically when conditions become crowded. Normally shy and solitary, the close proximity of other locusts turns them into party animals. In the case of the African desert locust, this is because the proximity stimulates them to produce the neuromodulator chemical serotonin, which not only makes them gregarious but also stimulates other nearby locusts to generate serotonin as well. The ensuing chain reaction soon has all the locusts in the vicinity seeking each other's company.

The locusts also become darker, stronger, and much more mobile. They start moving en masse, first on the ground and then in the air, gathering other locusts as they go and forming dense swarms that can eventually cover an area of up to 500 square miles and contain a hundred billion locusts, each eating its own weight in food each day over their lifetime of two months or so.

Descriptions of such locust plagues appear in the Qur'an, the Bible, the Torah, and other ancient texts, and modern plagues affect the livelihoods of 10 percent of the world's population. It is little wonder that scientists want to understand what it is that drives the locusts to mass together and travel in such huge numbers. The behaviors they have uncovered give us vital clues about the self-organization of other animal groups, from social insects to human crowds.*

When individual locusts first start moving, they are still in their juvenile, wingless form. At first their movements are more or less random, but as the population density increases, their directions of movement become more and more aligned. When the population density becomes very high (around seven locusts per square foot), a dramatic and rapid transition occurs: the still somewhat disordered movement of individuals within the group changes to highly aligned marching.

Rather similar transitions happen in human crowds. At low densities, the movement of individuals can be likened to the random movement of molecules in a gas, as engineer Roy Henderson discovered when he monitored the movements of college students on a campus and children on a playground. In both cases, he found that the movements fit an equation (called the Maxwell-Boltzmann distribution) that describes the distribution of speeds among gas molecules. When he applied the theory to the observed movements of students and children, he found that their distribution of speeds followed the same pattern. The only difference between the students and the children was that the children had much more energy and consequently moved at much higher average speeds.

Video studies of pedestrians show that their movements have a similar random component, although an overall direction is super-

* Luckily for us, of the twelve thousand known grasshopper species, fewer than twenty are locusts, although there are species of locust endemic to every continent except Antarctica.

posed on their movements by the desire to reach a goal. When the pedestrian density reaches a critical value, however, spontaneously self-organized rivers of pedestrians start to form and flow past each other, with everyone in a particular river walking at the same speed, just like marching locusts.

How does such self-organization occur? Are the basic mechanisms the same for locusts and for people? Can the collective behavior of locusts and other insects tell us anything about the behavior of human crowds? Over the course of the next four chapters I give answers to these questions, beginning here with the fundamental one: what are the forces that produce swarm behavior?

In the case of marching locusts, one of those forces is the simple desire not to be eaten by the locust behind! Marching locusts are in search of food, and the locust in front provides a tasty temptation. The way to avoid being eaten is to keep marching and to keep your distance, just as the way to avoid being pushed from behind in a human crowd is to keep moving steadily forward.

But keep-your-distance is not in itself sufficient to explain the self-organized synchrony of a group of marching locusts. If that were all there was to it, the group would simply disperse. There must be a balancing force to hold the group together.

That force is provided by the serotonin-induced drive for company, which increases disproportionately with the number of similarly inclined locusts nearby. It is, in other words, nonlinear, and positive feedback is thrown in as more and more locusts are recruited, increasing the gregarious drive of those already in the group—just the conditions that are needed for the emergence of complex collective behavior.

To understand how that collective behavior emerges, we need to turn to computer simulation. The first such simulation (produced in 1986 by animator Craig Reynolds) used small triangular objects called "boids." The original animation is still worth a look. It laid the foundation for all of our subsequent advances in understanding complex collective behavior.

Boids

Reynolds' boids are small isosceles triangles. They wheel, dive, and disappear into the distance in a manner highly reminiscent of flocks of real birds. The audience that experienced their first public presentation, at a conference on "artificial life," was particularly impressed by the way in which the flock split into subflocks to go around a pole (a circle on the screen) and then unified itself again on the other side. They were even more impressed when one boid crashed into the pole, fluttered for a while as though stunned, and then flew on to rejoin the flock.

Such lifelike behavior would seem to require very complicated, very sophisticated programming. But in fact the program is quite short, and the individual boids follow just three simple rules:

- Avoid bumping into other individuals.
- Move in the average direction that those closest to you are heading.
- Move toward the average position of those closest to you.

These can be more succinctly described as:

- *Avoidance* (*separation*)
- *Alignment*
- *Attraction (cohesion)*

Next time you find yourself in a crowd at an airport, a train station, or a football game, take some time to watch those around you as they walk. You will usually find that most people are obeying the same three rules.

Reynolds' goal was to demonstrate that lifelike collective behaviors can emerge from simple interactions between individuals. Although he did not know it at the time, the three rules he used corresponded to the empirical rules discovered by Brian Partridge in his studies of fish schools. Partridge didn't mention avoidance (probably taking it as obvious), but the other two rules he identified are equivalent to Reynolds' rules of alignment and attraction. The optimum way for all fish to maintain pace with those alongside them and simultaneously following those fish in front is to move in the average direction of the nearest individual fish and toward their average position (concomitant with not actually bumping into them).

Reynolds' original model was taken up enthusiastically by the computer animation industry, where it is still used today. Its value to that industry is undoubted, but its deeper worth lies in the help it continues to give us as we unravel the secrets of collective behavior, such as that of locusts marching in synchrony.

Locust Logic

Computer simulations have shown that synchrony emerges because each locust acts as a self-propelled particle whose velocity (i.e., speed *and* direction) is determined by those of its neighbors according to a specific built-in rule. This sounds like one rule rather than three, but a closer look reveals that this single rule can be decomposed into three rules that are similar to those proposed by Reynolds: follow the locusts in front, keep pace with the locusts alongside, and keep your distance from the ones behind. When all locusts in the group obey the same rules, the result is synchrony. (I show in chapter 4 that a similar synchrony can emerge in dense human crowds.)

When locusts develop wings and start to fly, things change. Now the whole sky is at their disposal, and they have more to fear from birds and other predators than from their fellow locusts. The direction the swarm takes is determined by the wind, but the urge to stick together is still strong, since flying with the group reduces the risk of predation on any individual. But when flying, locusts need more space, because a midair crash could damage their delicate wings, leaving them on the ground in an area where every vestige of food has already been consumed by the swarm.

The new balance of forces still reflects in Reynolds' three rules, but the relative importance of the rules is different. The last two rules become relatively weak (although still strong enough to keep the swarm together), while the avoidance rule becomes stronger.

Implementing the avoidance rule starts with increased sensitivity to the presence of moving objects, especially to those coming from the side. The early *Star Wars* experiments showed that locusts flinched mentally when they noticed such objects approaching. Later experiments, in which the locusts were allowed to fly freely (although tethered to a length of cotton) showed that their response to an object coming from the side was to close their wings and go into a brief diving glide. This strategy gives them the best chance of avoiding a collision, and of protecting their wings if there is a collision.

We adopt a rather similar strategy when walking in dense crowds. Instead of folding our wings, we keep our arms close to our sides. Instead of going into a diving glide, we shorten our steps or even stop moving. The overall effect, as with locusts, is to strengthen the avoidance rule as much as possible.

This and other minor modifications of Reynolds' three rules are sufficient to explain many aspects of swarm behavior, but no simple modification of the rules is sufficient to explain the emergence of true swarm intelligence. Reynolds' rules explain how a group can collectively respond to external circumstances, but swarm intelligence needs something more—the ability to learn. This requires some additional form of communication within the group—the sort of communication that is shown, for example, by bees.

Bee Logic

Individual bees in swarms follow the basic rules of avoidance, alignment, and attraction, but the swarm as a whole has something that locust swarms don't—an ability to fly directly to a target that has been identified by scouts. The way the swarm does this provides the first clue to the processes by which swarm intelligence emerges.

"Well," you might think, "it's pretty obvious how they find the target. They use the well-known waggle dance. It's the method that bee scouts use to tell the others where something is, such as a food source or a site for a new home. The scouts dance like teenagers in a disco, waggling their abdomens while moving in a tight figure eight. The overall direction of the dance points in the direction of the target, and the speed of the waggling tells how far away it is.

Unfortunately this explanation doesn't provide a full answer. The dance is performed in a hive that is almost as dark as some discos, so only those bees nearby (about 5 percent of the total) see the dance. The majority doesn't see it, so most bees start flying in complete ignorance. Those that have seen the dance aren't even out in front, showing the others the way. They are in the middle of the swarm, flying with the rest. So how does the swarm find the target?

There seem to be two main possibilities: (1) the bees who know where the target is might emit a pheromone, and (2) those bees may behave in a way that flags them as the leaders. To check out the first possibility, scientists covered the Nasonov glands (the ones that emit pheromones) of each bee in a swarm with a spot of paint. They found that the swarm still flew straight to the target, which disproved the pheromone hypothesis.

The swarm-following scientist Martin Lindauer discovered a clue that pointed to the possibility of leaders when he looked closely at swarms flying overhead. (Presumably he was running on flat ground.

If it had been me, I would have tripped over the nearest tree root and fallen flat on my face.) He noticed that a few bees were flying much faster than the others in the swarm, and that they seemed to be flying in the direction of the target.

Fifty years later, other scientists confirmed his observation by photographing a swarm from below, leaving the camera aperture open for a short length of time so that individual bees appeared as dark tracks against the sky. Most of the tracks were short and curved, but a few tracks were longer (indicating that the bees were flying faster), and also straighter, with the lines pointing toward the target.

The bees that produce the straight tracks have been evocatively named "streakers." It seems at first that these would be the bees that know where the target is and that their behavior is intended to guide the other bees. It still remains to be proved whether the streakers are those that have received information from the scouts, but computer simulations of bee swarms have produced a stunning surprise—*it doesn't matter.*

Simulations have revealed that the knowledgeable bees do not need to identify or advertise themselves to the rest of the swarm to lead it successfully. Just a few informed individuals can lead a much larger group of uninformed individuals simply by moving faster and in the appropriate direction. Guidance is achieved by way of a cascade effect, in which uninformed individuals align their directions with those of their neighbors. Even if only a few bees know their way, Reynolds' three rules—avoidance, alignment, and attraction—ensure that the whole swarm moves in the direction that those knowledgeable bees take.

Leadership by these few individuals arises, according to the computer modelers, "simply as a function of information differences between informed and uninformed individuals." In other words, it needs only a few anonymous individuals who have a definite goal in mind, and definite knowledge of how to reach it, for the rest of the group to follow them to that goal, unaware that they are following. The only requirements are that the other individuals have a conscious or unconscious desire to stay with the group and that they do not have conflicting goals.

The purposeful movement of bee swarms, in other words, is an example of an emergent complex behavior that arises from simple local interactions guided by appropriate rules.

Japanese scientists have already taken advantage of this discovery to design robots that will swarm around a human leader and follow the leader happily across a factory floor as they are led to perform a task. The robots have no goals of their own, just a built-in desire to stay with the group, using only Reynolds' three swarm rules to stay together and follow the leader.

Could rules that apply to robots also apply to us? Surely if we were in a group, we wouldn't blindly follow unidentified "leaders" to a goal that only the leaders knew about?

Oh yes we would.

Invisible Leaders

Volunteer groups of university students were asked to participate in an experiment in which they were instructed to walk randomly in a circular room that had labels with the letters A to J distributed uniformly around the wall. The students were instructed to walk at normal speed, and not to stop until they were told to. They were allowed to walk anywhere in the space, required to stay within arm's length of at least one other person, and forbidden to talk or gesture. This way, they met the swarm criteria of staying with the group but not having any particular goal in mind.

A few of the students were given an additional secret instruction: go to a specific label, but without leaving the group. By the time the students were told to stop walking, most of them had ended up near the same label. They were led there, but they did not know that they had been led.

We tend to think of leaders as being highly visible and needing specific qualities in order to lead effectively. Leadership guru John C. Maxwell, author of the best-selling books *The 21 Irrefutable Laws of Leadership* and *The 21 Indispensable Qualities of a Leader*, lists qualities such as charisma, relationship, and vision as being essential.

The above experiments show, however, that there is another possibility: we can lead a group simply by having a goal, so long as the others in the group do not have different goals.

Leading from within is, of course, a well-known strategy, encapsulated in the phrase "the power behind the throne" or the term *éminence grise*. This sort of leading, though, is not what the experiments or the simulations were about. Persons such as Dick Cheney, Edith Wilson, and Cardinal Wolsey have exercised an influence that was not always visible from the outside; but George W. Bush, Woodrow Wilson, and Henry VIII—their respective puppets—certainly knew who was pulling the strings. What the computer models predict, and what experiments show, is that members of a group can be totally unrecognized as leaders *by those whom they are leading.*

Furthermore, computer simulations have shown that "the larger the group, the smaller the proportion of informed individuals needed to guide the group with a given accuracy." In the case of the students, the group needed only ten informed people out of two hundred (just 5 percent of the group) to have a 90 percent chance of success in leading the rest of the group to the target.

Sometimes the target doesn't even need to be a real one. In 1969, the famous social psychologist Stanley Milgram arranged for groups of people to stand in a New Haven, Connecticut, street and stare up at a sixth-floor window, an experiment that has become a classic. With just one person staring up, 40 percent of passersby stopped to stare with them. With two people, the proportion rose to 60 percent, and with five it was up to 90 percent. His results conform beautifully with later discoveries about invisible leaders.

Maybe Connecticutians are unduly gullible, but when I repeated the experiment on a busy street in Sydney, I found that Australians are equally gullible, or at least biddable. I made my leaders even less visible, having them melt away when the crowd became large enough, leaving a crowd that stared by itself.

On a more serious note, the presence of a few knowledgeable individuals in a swarm can make a world of difference to its performance. Without such informed individuals, the group can only react to external circumstances, as fish do when their school reacts as a unit to the approach of a predator, or as locusts do when they fly as a group in the direction of the prevailing wind. Swarm intelligence in the absence of individual knowledge and goals keeps a group together and allows it to react to situations, but it is difficult, if not impossible, for the group to be proactive.

Bee logic changes all that. The Grammy Award–winning Orpheus Chamber Orchestra provides a real-life example. Audiences at Carnegie Hall who have not previously seen the orchestra can be surprised to see the black-clad members take their place on stage and begin to play—without a conductor. The orchestra, also the winner of a WorldBlu—a "Worldwide Award for the Most Democratic Workplaces"—appears to produce its beautifully coordinated sounds by democracy alone. How does this work?

The orchestra does it by using invisible leaders. The music does not degenerate into an anarchic mess because a core of six out of the thirty-one players sets the musical agenda for each piece. The leaders are not invisible only to us in the audience. They are also effectively invisible to the other players during the performance. Those players are aware of who the leaders are, but they are not consciously watching them in the same way that they would watch and obey a conductor. They have, however, set aside their own agendas so that they are free to be swept along by those few in the group who do have a specific agenda.

The idea of an invisible leader working within a group is as ancient as civilization itself. According to a Chinese proverb often attributed to Laozi, the founder of Taoism, "a leader is best when people barely know he exists . . . when his work is done, his aim fulfilled, they will say, 'We did it ourselves.'"

What *is* new here is the proof, both theoretical and practical, that a leader (or group of leaders) can guide a group toward an objective from within and never be recognized. This suggests a rule that we can use as individuals within a group: *Lead from the inside (if possible with a coterie of like-minded friends or colleagues), but take care not to let other members of the group know what you are doing. Just head in the direction that you want to go, and leave it to the rules of the swarm to do the rest.*

The process works in groups of individuals who have an innate or learned tendency to follow the example of those nearby. Just a few individuals taking the lead instead of copying is sufficient to induce a chain reaction of copying, and soon the whole group will be following their example. Any deviations will quickly be brought into line by negative feedback, physical and social pressures conspiring to push deviant individuals back into moving with the rest. The more the deviation, the stronger the pressure.

I asked Jens Krause, the supervisor of the original experimental study with the students in the circular room, whether he knew of any real-life examples of leading from within. He gave me one from his own experience. "Recently I got off the plane in Rome at midnight," he said,

> and the airline stewards provided no help in directing us to the terminal. It was dark, the passengers didn't know each other, nobody talked and most people looked utterly clueless. But sud-

> denly two of them walked off purposefully in a particular direction and the group self-organized into following them [in a chain reaction where a few followed the first two, and then a few more followed each of them, and so on]. Sure enough, they guided us to the right terminal.
>
> When the [experimental] study was published, we were contacted by historians of warfare who pointed out that the leadership of small groups can engage a whole army. Police officers pointed out that they try to remove a small proportion of troublemakers during demonstrations or town fights to control whole crowds. At conferences it often happens that scientists stand in small groups talking to each other and not realizing that it is time for the next thing on the agenda. However, it is enough for a few people to start walking, and once they initiate a direction, most people follow (often while still talking), not anticipating where they are going or what the next activity will be—they realize this only after they arrive.

These processes sound obvious, but the ways in which individual knowledge and behaviors can influence a group depend on a subtle dynamic interplay of positive feedback, negative feedback, and cascading chain reactions. Locust and bee logics provide important clues as to how these processes can interact to produce swarm intelligence, but these are just the first pieces in the puzzle. As I show in the next chapter, many more pieces have been discovered through the study of another social insect—the ant.

THREE

Ant Logic

"Go to the ant, thou sluggard; consider her ways, and be wise." So says the biblical proverb, and modern scientists have been following its sage advice to learn from the ants. In doing so they have learned many lessons about the evolution of complexity and have even been able to produce a computerized version of the ants' approach to problem solving.

Ants face difficult decisions in their lives, not least in their choice of routes to a food source, where it is important to establish the shortest route so that they waste as little energy as possible when carrying the food back to the nest. Judging by the ant trails in my garden, they seem to do it pretty efficiently. The trails are invariably straight, representing the shortest distance between two points. Ants can distinguish objects up to three feet away (depending on the size of the ant), but their trails can be many feet long, with the food source hidden from view by intervening rocks, leaves, and sticks. How do the ants establish such wonderfully straight trails?

Experiments on a laboratory colony of Argentine ants (*Iridomyrex humilis*) provided the answer. Researchers at the Unit of Behavioral Ecology in the University of Brussels set up a bridge between the colony and the food source. The bridge was split in the middle to offer a choice between two curved routes, one twice as long as the other. The first ants to leave the colony in search of food chose one or the other branch at random, but within a few minutes practically the whole colony had discovered the shortest route, just as we rapidly discover a shortcut that provides a quicker way to get from home to the office.

"Finding the shortest route," said the scientists, "is extremely important, not only for Roman road builders, thirsty rugbymen, and applied mathematicians working on this very problem, but also for any animal [including humans] that must move regularly between different points." They discovered that the ants were finding the shortest routes not by looking at their watches to check the time but by using chemical signaling compounds, pheromones, which they laid down as they traveled so that other ants could follow their trail. But how could these signaling compounds help them to find the shortest route?

The reason, as Sherlock Holmes once said, is very obvious—once you think of it. The first ants to return to the nest after foraging are those that happen to have chosen the shortest path. They will have laid down pheromones to mark the trail, and this trail will be followed by other ants, who are genetically programmed to "follow the pheromone." By the time that ants using the longer trail have returned, there will be more ants using the shorter trail, and laying down yet more pheromone. Furthermore, ants that went out by the longer trail but returned by the shorter trail will also add pheromone to the shorter trail. In the end, the shorter trail will have an overwhelmingly higher concentration of pheromone, so it will be preferred by most of the ants.

The ants use the selective reinforcement of a good solution to find the most efficient route to their food source. We use a similar process to find shortcuts when we're driving. When someone finds a shortcut, a few of other people might notice others leaving or rejoining the main route and might follow them on the suspicion that they have found a shortcut. Each of those might be noticed by a few more people, and so on—and the cascade amplifies. This process of positive feedback ensures that soon everyone knows about the route. We don't even need pheromones to give us the message—just observation.

In a computerized version of ant logic known as "ant colony optimization," positive feedback is similarly beneficial. Let's say that a programmer is faced with the problem of planning a bus tour between a number of cities. How can she determine the shortest route, or the quickest one if roads with different speed limits are involved?

It sounds easy to solve, but an exact mathematical solution to this class of problem (known as the "traveling salesman problem") has eluded mathematicians for centuries—and it continues to elude them. The problem is so important, both theoretically and for practical applications, that it even has its own website (http://www.tsp.gatech.edu/), which contains much information on its history and applications.

One way to tackle the problem, with the aid of modern computers, is simply to measure the length of travel time for all of the possible routes and then choose the best from the list. This might work when there are only a few cities involved, but the calculations can get out of hand rapidly. To calculate the optimal route that Ulysses might have taken between the sixteen cities mentioned in *The Odyssey*, for example, requires the evaluation of 653,837,184,000 possible routes by which he might have gone out and returned home. That adds up to something like ten thousand billion calculations, which takes some doing, even for a modern computer.

Ants do it a different way, using the positive feedback principle to obtain a good approximate solution. The virtual insects in computer simulations are let loose in an imaginary world of the sixteen cities with instructions to visit every city before returning home. The cities are joined by imaginary lines (called "links") in such a way that each city is connected to every other one, with the distance between each pair of cities being represented by the length of the line.

Now comes the clever bit. When a virtual ant returns home, it remembers how far it has traveled and attaches a number (the equivalent of a pheromone) to each link that reflects the total length of the journey. Each link gets the same number, and the shorter the journey, the higher the number. As more and more "ants" travel through the network, those links that belong to the shortest journeys accumulate higher and higher total numbers (equivalent to higher concentrations of pheromone). The numbers increase even more because the following ants are instructed, when confronted with a choice of links, to show some preference for the one that already has the highest number associated with it.

And now for the really clever bit. The size of the numbers gradually diminishes with time in a programmed countdown that corresponds to the slow evaporation of the pheromone trail, which is what happens in the real world. The effect of this is a disproportionate reduction of the numbers for inefficient links (equivalent to the gradual de-emphasis of trails that are used less because of pheromone evaporation), so that the most efficient links stand out more clearly. Soon, the most efficient route (or one that is very close to the most efficient) stands out for all to see. Ant colony optimization has done its job.

It is doing many such jobs now, especially in the telecommunications industry, where the traveling salesman problem refers to routing messages through complicated networks in the most efficient manner. The messages themselves become the ants, recording their progress and marking the trail accordingly.

Could we use a similar procedure to resolve traveling and networking problems in our own lives? Robert J. Dillon, one of the original Central Park commissioners, had one idea when he suggested in 1856 that the planning of pathways in the park should be postponed until New York City pedestrians had established them by habit, with the more deeply marked paths corresponding to those that were most used and therefore most efficient.

Dillon did not get his way, but recent research by German traffic engineer Dirk Helbing and his colleagues has shown that Dillon's solution, a neat example of ant colony optimization as practiced in human society, would have been a good one. Helbing and his colleagues have photographed and analyzed many such paths, and when I asked him about the validity of Dillon's approach he replied,

> If people use trails frequently enough, direct connections between origins and destinations are likely to result. If the frequency of usage would not be high enough to maintain direct trails in competition with the regeneration of the natural ground, however, pedestrians with different destinations form commonly used trail segments and are ready to make compromises. In that case they accept detours of up to 25 percent, but the resulting trail system will usually provide an efficient and fair solution that optimizes the walking comfort and minimizes detours within the limits of how many trails can be maintained by the usage frequency.

When ant colony optimization is available to us, it seems that we use it spontaneously, producing solutions that are reasonably close to optimum. Getting the best solutions, however, requires very carefully setting conditions, as has happened with the communal website Digg.com, which allows its users to submit news stories they find as they browse the Internet.

New submissions appear on a repository page called "Upcoming Stories." When other members read the story, they can add a digg point to it if they find it interesting. If a submission fails to receive enough diggs within a certain time, it is removed. If it earns a critical number of diggs quickly enough, it jumps to the front page, where it may continue to receive more diggs.

This process of positive feedback is counterbalanced by the fact that the novelty of stories decays with time. The effect is similar to pheromone trail evaporation and link number decline: the stories receive less and less notice and fewer and fewer diggs. Eventually they disappear off the front page, to be replaced by newer and more interesting stories that have come to the fore.

The "Letters to the Editor" section of my local newspaper follows a similar pattern. If a topic attracts enough letters, more letters are likely to pour in, and editors seem to be more inclined to print letters on that topic. With time, boredom sets in, the flood of letters slows to a trickle, and editors may announce a stop to the correspondence on that issue.

Positive feedback can be very useful for bringing attention to an issue and keeping it on the public agenda. This just needs a bit more thought and planning than many community action groups seem to give. Taking a lesson from ant colony optimization, the best strategy is not to fire off a whole slew of letters at once and then leave it at that, but to plan for members of a group to keep up a steady flow of letters on different aspects of the issue. This strategy is equivalent to continually adding to the concentration of a pheromone before it has time to evaporate, and it suggests another rule: *When trying to bring an issue to the notice of a group or the public as a whole, don't be a one-hit wonder; plan to bring different aspects to the fore in succession over time.*

Ant colony optimization is useful for suggesting analogous strategies, but its lessons are not the only ones we can learn from ant logic. Closer to our everyday experience is a modification of ant colony optimization called "ant colony routing" In this process the antlike agents that inhabit a virtual computer world learn from experience what the shortest and fastest routes are. The next time they are called upon to perform, they use their memory of the routes rather than relying on signs left by previous users. If the routes are those in a communications network, for example, the agents remember the parts of the network that are most likely to become congested and seek new routes, just as we do when we have a choice of a number of routes between home and work. The new routes eventually become congested in their turn, but ant colony routing can cope with these dynamic changes in a way that standard ant colony optimization does not.

The ultimate application of ant logic to problem solving has come in the form particle swarm optimization, a combination of locust, bee, and ant logics that no insect could have come up with, which emerged from the fertile minds of Russell Eberhart and Jim Kennedy (from the Purdue School of Engineering and Technology at Indiana University and the U.S. Bureau of Labor Statistics respectively) as the culmination of a search for a form of computerized swarm intelligence with the broadest possible problem-solving ability.

The way it works is a bit like sitting for an exam in which cheating is allowed. Each candidate writes down his best answer but is allowed to look over the shoulders of those near him and to modify his answer if he thinks that someone else's is better.

That's not the end of it, though, because the student next door may come up with still a better answer after looking at the answers of the students near *her*. Then the first student can produce a further improvement by copying *that* answer. Over time, the whole class will eventually converge on the real best answer through this process of positive feedback.

The lesson is this: if you see people doing something a better way, then copy them. It is one that I was reminded of when my adult son was helping me drag dead branches along our lengthy driveway to be put out on the street for mulching. I noticed that he was gradually getting ahead of me as the pile grew, and I couldn't figure out why until I realized that the pile was on the left, and he was dragging the branches under his left arm (and so could drop them straight on the pile) while I was carrying them under my stronger right, but had to then put them down, walk round them, and pick them up again to place them on the pile.

The lesson is especially effective in larger groups, in which best practices can rapidly spread throughout by way of positive feedback and repeated learning from those nearby. Eberhart and Kennedy produced a computer analog of this process by replacing the virtual ants of ant colony optimization with particles, which in this case were guesses at the solution to a problem. (So a particle could be an equation, for example, or a set of instructions, depending on the problem.) A swarm of such particles is then allowed to fly through the space of the problem, remembering how well they have done but also noticing how well nearby particles are doing.

The particles follow similar rules to those that Reynolds used for his boids. A particle's movement is governed by the balance of two forces, one attracting it toward the fittest location that it has so far discovered for itself, and the other attracting it to the best location uncovered by its neighbors. It's easier than it looks, as can be seen by viewing one of the many computer visualizations that are available, such as that at Project Computing (http://www.projectcomputing.com/resources/psovis/index.html), where a swarm of particles attempts to find the highest peak in a virtual mountain landscape.

Particle swarms are particularly good at detecting abrupt changes in their environment, such as peaks, troughs, edges, or sudden movements. The ability to detect peaks and troughs has even made particle swarm optimization a useful adjunct for investment decision making. It has been adopted as a tool for the analysis of MRI scans and satellite images and for automatic cropping of digital photographs because of its ability to detect edges, and its facility for movement perception makes it useful for detecting intruders, tracking elephant migrations, and analyzing tremors in the diagnosis of Parkinson's disease.

All of these applications require the use of powerful computers, but this does not mean that computerized swarm intelligence has taken over our decision making. Rather, it provides new and exciting opportunities for innovative approaches to problems, such as that taken by UPS when it combined years of accumulated company know-how with the route-planning capabilities of new software to plan as many right turns into its routes as possible.

The reasoning was obvious. Almost every left turn involves crossing an oncoming lane of traffic, which means a potential wait and consequent loss of time, as well as a greater risk of an accident. It was the accumulated experience, though, and the process of drivers learning from each other in a way similar to ant colony routing, that convinced the firm that more right turns really would save time. This communication among agents, together with new package flow procedures (also designed with the help of ant colony logic!), resulted in substantial fuel savings: three million gallons in 2006 alone.

If UPS can do it, so can we. UPS used the swarm intelligence of their drivers to come up with the strategy. We can take advantage of that swarm intelligence by copying it. So here's another rule: *In complicated journeys across a city, choose a route that incorporates a high proportion of right turns.*

The human-computer interface permits a whole new approach to the development of human swarm intelligence. Astronomers are using it to coordinate their activities in the hunt for supernovae. eBay shoppers are unconsciously using it to maintain quality control over transactions through the rating system, which works through a combination of positive, negative, and neutral feedbacks. Hybrid systems, consisting of hardware, software, and humans, are also beginning to emerge. They allow simple local interactions between neighboring individuals (human and computer) to produce complex swarm intelligence, and an improved performance for the group as a whole. Tests on volunteer swarms in the U.S. Navy have shown that this sort of system can work well when it comes to cargo movement on navy ships, and there are many other potential applications in the pipeline.

One of the most dramatic uses of the human-computer interface is in the production of a "smart mob." Smart mobs (also known as "flash mobs") are groups of people who use cell phones or other modern communications media to coordinate their activities. Such communication leads to swarm intelligence because communication on a one-to-one basis enhances the performance of the group as a whole, with no obvious leader. The communication is particularly efficient because of exponentially increasing network links between members of the group over time.

Smart mobs can be very frightening, especially to authorities. Protestors in the 2001 demonstrations that helped to overthrow President Joseph Estrada of the Philippines became a self-organized group linked by text messages. The 2005 civil unrest in France, the 2006 student protests in Chile, and the 2008 Wild Strawberry student movement in Taiwan (where a group 500-strong materialized in front of the National Parliament building overnight to protest curbs to freedom of expression) were similarly self-organized. More recently, the Twitter network has been used to some effect to coordinate protests in Iran against the results of presidential election. It should be noted, however, that this sort of networking has the weakness that it is "chaotic, subjective, and totally unverifiable," and it is also impossible to authenticate sources.

Smart mobs do not always have to be concerned with protest, however. The technology that empowers them, which can lead to "smart mob rule," has the potential to help all of us to enhance our day-to-day group performance. The Twitter social networking and microblogging service, for example, is now used regularly by politicians and celebrities as well as families, teenagers, and other social groups to keep in real-time contact with what everyone in the group is doing.

It need not stop at person-to-person communication. These days we can even communicate with our refrigerators and washing machines. Evolutionary biologist Simon Garnier, a specialist in insect swarm intelligence, has waxed lyrical about the potential for this sort of communication in the evolution of human uses for swarm intelligence: "We have no doubt that more practical applications of Swarm Intelligence will continue to emerge. In a world where a chip will soon be embedded into every object, from envelopes to trash cans to heads of lettuce, control algorithms will have to be invented to let all these 'dumb' pieces of silicon communicate [with each other and with us]."

We don't have to become cyborgs to use aspects of swarm intelligence in our daily lives, though. Nor do we have to become ants, although our brains themselves use the distributed logic of the ant colony. As I show in the next chapter, most of the time, all we need for human swarm intelligence to emerge is an ability to use straightforward human logic, and in some cases also use the much rarer ability to recognize the limitations of that logic.

FOUR

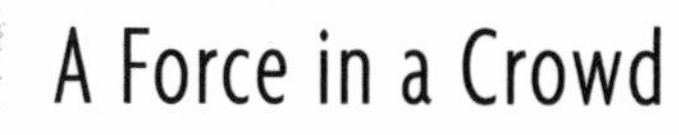

A Force in a Crowd

A friend of mine who uses a wheelchair has an unusual solution for coping with crowded conditions. He is a well-respected scientist with a Ph.D., but when he gets sufficiently irritated with people blocking his path on a crowded street, he has been known to adopt a goony expression and tap on their backs, or even tug at their clothing. Most of them back away hastily and apologetically, and as he goes past he turns to them with a normal expression and says politely, "Thank you very much."

The English humorist Jerome K. Jerome discovered another way to make space in a crowd when he boarded a packed train carrying a bag full of moldy cheese. His compartment emptied quickly after a fellow passenger who looked like an undertaker said that the smell reminded him of a mortuary!

Putting such extreme solutions aside, what are the best strategies to adopt in crowd situations? Should we take our lessons from locusts, bees, and ants, or does our ability to foresee the consequences of our decisions make a difference in how we should act as individuals in a human swarm?

Scientists have made remarkable progress in the past ten years in coming up with answers to these questions. That progress has largely arisen from the recognition that crowd movements involve a combination of involuntary and voluntary forces.

The scientists who study crowd dynamics call the involuntary forces "physical forces." They reserve the term for those forces that we experience when we are pushed from behind, when we bump into other people, and when we find ourselves trapped against a wall or other immovable object. The forces that we generate voluntarily to help us achieve our goals for movement in the crowd are called "social forces." If we want to head in a certain direction, for example, we use our legs to generate a social force that pushes us in that direction. If we want to stay close to family and friends, we continuously vary the magnitude and direction of the chosen force so as to keep near them. If we want to avoid bumping into other people, we move sideways to dodge the encounter.

The term "social force" can be misleading. It does not mean the sort of emotional reaction that caused people to jump aside when confronted by my friend in the wheelchair. It means the actual force that they exert when they push on the ground with their feet to make the move. That force came from a physical response to a social interaction, which is why it is called a "social force."

When conditions are not too crowded, the combination of forces that lead to staying together yet avoiding each other can be interpreted in accordance with Craig Reynolds' three rules for boids: avoidance, alignment, and attraction. We avoid colliding with others by aligning our movement with theirs, and we stay with the crowd because it occupies the space through which we want to travel.

The cumulative effect of both types of forces on our movements can be described mathematically by adapting the three laws of motion that Sir Isaac Newton famously propounded hundreds of years ago. Put briefly, they are:

1. Everything keeps moving in a straight line at a constant speed (which may be zero) unless it is acted on by an external force.
2. When a force does act on a body, the body is accelerated (which can mean a change in speed or direction or both) in proportion to the force, and in inverse proportion to the body's mass.
3. For every action there is an equal and opposite reaction.

Scientists still use these laws to predict the movement of spacecraft, the acceleration of race cars, and the trajectories of balls on a pool table. Now they are also using them to calculate our movements when we are pushed and buffeted as we make our way through a crowd. The beauty of describing our involuntary and voluntary movements as the results of physical and social forces, respectively, is that we can use simple mathematics to add up all the forces and then calculate their net effect on our movements directly using Newton's Second Law of Motion.

Newton's Second Law says that an external force accelerates us in the direction of the force and that our acceleration is proportional to the force. Put mathematically:

$$\text{force} = \text{mass} \times \text{acceleration}$$

So a force that is twice as strong will accelerate us by twice as much. A closer look at the equation reveals a second fact that we also know from experience: the heavier a body, the harder it is to accelerate. So if we need to push our way through a crowd, our best strategy is to push past only the lighter people. (As we will see, though, there are often other ways to achieve the same objective, and a pushing strategy should be used only in exceptional circumstances. By avoiding that strategy we also avoid the likelihood of a punch in the eye.)

The first people to think of bundling physical and social forces together in this way were pioneering mathematical sociologist Dirk Helbing and his team of international colleagues. Their subjects for a computer simulation were members of a virtual soccer crowd—cylindrical individuals weighing in at around 180 pounds with shoulder widths between 20 and 28 inches, which seems pretty reasonable for a group of adults in heavy coats attending a soccer game on a cold winter's day.

This group of individuals didn't get as far as watching the match, however. Instead, the scientists marched them down corridors, around pillars, and through narrow exits—just to see what would happen. Being the inhabitants of a virtual computer world, they were in no position to object; they were compelled to behave according to the social forces that they were endowed with and the physical forces that they experienced. Their behavior in response to these forces told the scientists a great deal about what happens in real crowds, and it conformed to what has subsequently been observed in video records of such crowds. These studies have led to huge improvement in the way that many potentially troublesome crowd situations are now handled.

The Individual in a Crowd

When it comes to our individual behavior in crowds, the computer models revealed that it does not pay to try to weave one's way through a large group of people. Under normal circumstances, both model and experience show that pedestrian crowds self-organize. No one tells the people in them to do it; their actions are just manifestations of complexity theory, which reveals how simple local rules lead to complex overall patterns. The pattern in this case is of pedestrian streams through standing crowds, which form in a way that is analogous to the way running water forms riverbeds.

Army ants do something similar when they organize themselves neatly into three-lane highways as they travel between their nest and a food source that may be some distance away. The ants leaving the nest occupy the margins of the highway, and those returning carry their prey down the center.

The ants are practically blind, but they manage to organize themselves by following pheromone trails laid down by ants that have previously traveled along the route, and by using two additional social forces. One of those, an addition to the basic avoidance rule, concerns what happens when two ants traveling in opposite directions meet head-on: both ants turn, but the ant traveling away from the nest turns faster.

The other rule is for individual ants to keep on going in the direction that they were already going after an encounter with another ant. Computer models have shown that these two rules are sufficient to account for the way in which ants that can hardly see each other form their highways.

We, too, form spontaneous lanes when we are walking in opposite directions in high pedestrian densities. The flow of people above a critical density of approximately one pedestrian for every two square feet of space breaks into interpenetrating streams moving in opposite directions.

The main difference between us and ants is that ants are programmed to follow the rules that lead to the formation of their highways while we have individual goals. Yet the outcomes are strikingly similar, and computer models have shown that the combined social forces of wanting to get to a goal and not wanting to bump into others cause us to form streams just as effectively as do the social forces experienced by army ants.

Flowing streams of pedestrians are obviously more efficient than individuals trying to work their way around each other as they walk in opposite directions. But can the computer models of this process help us personally to work out the best rules to follow when we are walking on a crowded street? If we could use it to make streams of pedestrians form for our benefit, that would be a good start.

The obvious way to do this is to increase the local pedestrian density. I decided to give it a try with a group of about twenty friends. We went out into a moderately crowded street and started walking. At first we spread out, but gradually we worked our way toward each other so as to box others in and thus increase the local density of pedestrians. As we did so—bingo! Just as the model predicted, the other pedestrians joined with us to form a column that thrust its way through the mob of pedestrians coming in the opposite direction in the manner of Moses parting the waters of the Red Sea.

What the experiment suggested was that *when you are working your way through a crowd with a group of friends, stick together closely and try to get as many strangers as possible to stick with you to promote the formation of a river flowing your way.*

The next time you're walking down a crowded street, you're probably going to notice such rivers of pedestrians. In bidirectional flows, these tend to separate into two unidirectional flows (so-called lanes). This segregation effect supports a smooth and efficient flow by minimizing friction through minimizing the number of interactions that cause deceleration. But when the density is too high or the crowd is too impatient, these lanes are destroyed and the crowd locks up. In one of the few experiments that I really regret having done, a group of friends and I deliberately tried to make this happen on a very crowded street. We were too successful by half, and I definitely don't recommend anyone else initiating a similar experiment. A crowd that had been moving fairly smoothly suddenly stopped dead. Fifteen minutes passed before it started moving again.

In less extreme circumstances, pedestrians who weave through a crowd have a similar effect to that of drivers who weave in and out of lanes of traffic. As shown by many traffic studies, the net effect of such driving behavior is to slow the traffic down without gaining any real individual advantage. Similarly, simulations of crowds show that if everyone in a crowd tries to move twice as fast, the net effect is to *halve* the rate of flow of the crowd.

I once had a competition with a friend on New Year's Eve in Sydney, Australia, when dense crowds of people were streaming toward vantage points to see fireworks. One favorite place to watch the display is the concourse of the Opera House at Bennelong Point, which has a view across the harbor toward the bridge from which the fireworks are launched. We set off to walk the three hundred yards from the ferry terminal; my friend tried to push through the crowd while I kept pace with the people instead. So that we wouldn't influence each other, we started off five minutes apart. I have to admit that he beat my time—by three seconds—hardly worth the extra effort.

The best way to navigate effectively in a crowd is to be aware of spontaneous crowd dynamics, going along with those dynamics rather than disrupting them. But what should we do when we come to a bottleneck, with people wanting to get through in both directions? Go with the flow, but be aware that the flow will now oscillate.

Crowd Self-Organization

Crowd self-organization is yet another example of complexity theory, self-organization, and collective intelligence. Crowds of pedestrians, who come from multiple directions and interact only with others nearby, self-organize so that they can pass through a bottleneck in the most efficient manner.

How do they do it? According to Helbing, it's all a matter of social forces:

> Once a pedestrian is able to pass through the narrowing, pedestrians with the same walking direction can easily follow.

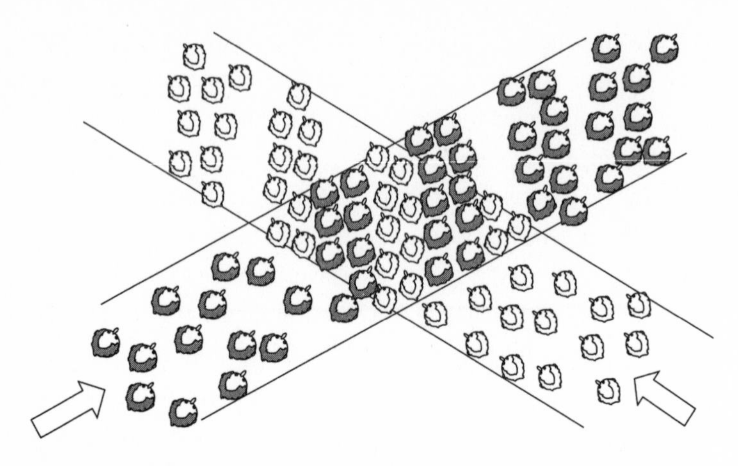

Schematic representation of intersecting pedestrian streams, with self-organized stripes forming perpendicular to the sum of the directional vectors of both streams. Redrawn from Dirk Helbing et al., "Self-Organized Pedestrian Crowd Dynamics: Experiments, Simulations, and Design Solutions," *Transportation Science* 39 (2005): 1–24.

> Hence, the number and "pressure" of waiting, "pushy" pedestrians on one side of the bottleneck becomes less than on the other side. This eventually decreases their chance to occupy the passage. Finally, the "pressure difference" is large enough to stop the flow and turn the passing direction at the bottleneck. This reverses the situation, and eventually the flow direction changes again, giving rise to oscillatory flow.

Self-organization can even occur when pedestrian flows cross each other. Just two social forces are involved, driven by the desire to keep moving forward and to avoid bumping into others. The net result is stripe formation (see figure above), in which pedestrians move forward *with* the stripes (so as to progress toward their destination) and sideways *within* the stripes (so as to avoid colliding with crossing pedestrians). It's a more complex version of the lanes into which ants spontaneously organize themselves, and, to my scientist's eyes, it has true inner beauty, as do so many of the complex patterns that emerge from simple rules.

All of these aspects of pedestrian flow in simulated medium-density crowds have been confirmed by video recordings of real crowds. In fact, the videos have been used to fine-tune the simulations so that they correspond as closely as possible to reality. Self-organization allows pedestrian flow to proceed in an efficient manner. Any effort to improve our individual situation in the crowd is likely to disrupt the self-organization and slow everybody down, including ourselves. We can enhance our prospects mainly by being aware of the nature of the flow so that we can reinforce it, rather than disrupt it.

Escaping the Crush

Things change when crowd densities are high, especially in confined spaces. Corridors become blocked, exits become jammed, and panic can set in. The physical analogy under these circumstances is not so much the flow of a liquid as it is trying to pour breakfast cereal out through a narrow opening in a box. Inevitably, a plug will form, the bits of cereal will jam the opening, and nothing will come out. If you shake the box, some bits will break free in a brief cascade, but as soon as you stop, the opening will become blocked again.

Simulations show that something very similar happens with crowds of people trying to make their way through a narrow exit. Under pressure from behind, the crowd density increases to produce "an irregular succession of arch-like blockings of the exit and avalanche-like bunches of leaving pedestrians when the arches break." The faster-is-slower effect dictates that the overall leaving speed becomes much more sluggish than it would be if everyone slowed down a bit and took their turn.

Even in the polite confines of a London theater, it appears that rushing is the order of the day. When I watched people emerging from a side entrance after a performance (knowing from previous experience that this is the route people take when they are in a hurry), they certainly emerged in bursts, rather than in a steady flow.

One way to overcome this effect is to subdivide the crowd into small enough blocks, but this is not something that an individual can readily do. If you are with a sufficiently large group, though, it could benefit the group as a whole to hang back a bit and let those in front make their way through the exit without being pressured from behind.

In the main, though, it is the responsibility of designers and architects to improve the geometry of entrances and exists so that passing through is easier. One of the more surprising conclusions from simulations of crowd movements through exits is that jamming can be initiated at places where an escape route *widens*; the crowd spreads out, leaving gaps into which other people can squeeze, so that jamming becomes really severe when the route narrows again. In 2005 Helbing and his colleagues came up with a clever solution: place columns asymmetrically in front of such exits. The existence of such columns can reduce the pressure at the bottleneck, and their asymmetrical placement can prevent an equilibrium of the forces of the flows from both sides and, thereby, mutual blockages. Simulations revealed that this solution works, and theater and stadium designers are now beginning to use it in practice.

If we are unfortunate enough to be caught in a crowd attempting to escape from a dangerous situation through multiple exits, what should we do? Should we go with the crowd, or should we go it alone in our efforts to escape?

Simulations of such situations suggest that we should do neither—or, rather, that we should do both. Our best chance of escape is to optimize our personal panic parameter, which is a measure of the extent to which we allow ourselves to be guided by the actions of the crowd rather than by our own initiative. If our panic parameter is 0, we search for an exit without bothering about what the crowd is doing. If its value is 1, we always follow the crowd. But that can lead to an inefficient use of emergency exits and alternative escape routes. Therefore, in a situation where we don't have any reliable information about escape routes, studies have shown that we do best when we operate with a panic parameter of 0.4—in other words, when we go with the crowd 60 percent of the time and use our own ideas and initiative 40 percent of the time. Just how we should mix it up would seem to depend on circumstances. My preference would be to do some personal searching first before linking up with the crowd (though still keeping an eye open for alternatives). I just hope, though, that I never find myself in a position to perform the experiment.

There are rational grounds for using such a mixed strategy, especially in conditions of bad visibility. Pure individualistic behavior (in the absence of prior information and knowledge) means that each person finds an exit only accidentally; pure herding behavior means that the entire crowd tends to move to the same exit, which can mean that some exits might not be used at all. When some individuals spend some of their time searching for alternative exits, all of the exits are likely to be discovered, with parts of the crowd following the individuals that have found them, so that all exits are efficiently used.

To use the theory effectively we need to face up to two basic facts about our real behavior in so-called panic situations: (1) we often fail to take the danger seriously until it is too late, and (2) even when we are aware of real danger, our response is often to seek out family and friends first before searching for exits or heading for the hills.

Our tendency to hang around and wait for confirmation of danger can be particularly disastrous. When my wife and I were in Sri Lanka just after the devastating tsunami in 2004, we passed through village after village hung with the white flags of mourning. Our driver was among those who had lost close friends. He told us that they might not have died if their natural curiosity had not impelled them to run out *toward* the ocean to find out what was going on when the waters receded rapidly—a signal (unbeknownst to them) that a tsunami was on its way.

When we later visited the west coast of India, our Indian friends told us of a different cultural tendency, one that saved many lives. Rather that running out to investigate, the locals headed for high ground as soon as the waters started to recede. Like the Sri Lankans, they didn't know what was coming, but they knew that the unusual movement of the water didn't fit with the expected order of things, and they wanted no part of it.

That's not the usual reaction, though; most of us are reluctant to accept and act upon warnings of a danger we do not perceive as immediate and personal. The recognition of danger can be delayed until it is too late, as sadly happened for many people in the tsunami. Our social forces for escape often do not kick in soon enough.

It seems that most of us have a built-in tendency to explain the unusual and abnormal in terms of the familiar and recognizable. The roar of an approaching tornado has often been mistaken for the sound of a passing train. Even when a tornado is apparent, it doesn't necessarily alarm some people. Writer Bill Bryson tells the story of how his grandfather in Iowa was woken up one night by a noise that sounded like "a billion hornets." He looked out the window, couldn't see anything, and went back to bed. When he got up the next morning, he was surprised to find that his car was standing in the open air. The entire garage had been taken by a passing tornado!

Less dramatically, but even more tragically, people have been known to die of carbon monoxide poisoning because they assigned the cause of their faintness to illness, and so failed to move away from the source of the leaking gas.

James Thurber's story of a whole town running from an imaginary flood may fit our stereotype of crowd behavior, but when the residents of Marysville and Yuba, California, were threatened with a major flood in 1955, 39 percent of those who received official warnings via the media "did not fully believe them." According to the official report on the subsequent disaster, the failure of many people to respond to the warnings was due to "lack of past experience with disasters, the delusion of personal invulnerability, the inability to adopt a new frame of reference so as to expect unusual events, dependency upon protecting authorities, and the willingness to seize upon reassuring communications or to deny or disregard communications predicting disaster."

In general, we do not react decisively enough to warnings of danger, even when the danger is extreme. After the 9/11 attack on the World Trade Center "as many as 83 percent judged the situation to be very serious in the first few minutes after the strike. Yet despite seeing the flames, smoke, or falling paper, only 55 percent of the survivors evacuated immediately; another 13 percent stopped to retrieve belongings, and 20 percent secured files and searched floors before evacuating." In another example, some residents who had escaped from a burning apartment building in Winnipeg, Manitoba, died, having gone back into the building to collect belongings, even though they could plainly see the smoke and flames.

The calmness we show in such conditions of crisis extends to our care for others, which seldom fits the stereotype promoted by the media, which suggests that we "panic, trample over each other, and lose all sense of concern for our fellow human beings." This is how the media interpreted events in Cincinnati, Ohio, in 1979 when eleven young people were killed in a crush at a concert by the Who. According to media descriptions, the tragedy occurred when crowd members stormed over others in their rush for good seats in the arena. One national columnist condemned them as barbarians who "stomped 11 persons to death [after] having numbed their brains on weeds, chemicals, and Southern Comfort." A local editor with literary aspirations referred to the "uncaring tread of the surging crowd."

The truth was otherwise, and it reflects well on our general behavior as human beings. An analysis of statements taken by police after the event showed that the overwhelming reaction of people to the crowd pressure was to attempt to help those nearby who seemed to be in trouble. One small teenager owed her life to four strangers who struggled to hold her off the ground after she had passed out. Thirty-seven of the thirty-eight interviews included descriptions of similar instances of altruistic behavior. Far from trampling on others, most people were much more concerned with helping them. It seems that the Good Samaritan is alive and well in most of us.

Other research on crowd situations has shown that we tend to seek out family and friends as a first priority. Social scientists call this the "social attachment" model of crowd behavior. It is not always the best objective strategy. It can waste valuable time, for example, which might be better used by heading straight for the nearest exit. In some circumstances, such as the aftermath of earthquakes, anxious relatives who are trying to be helpful can get in the way of professional large-scale rescue operations. The forces of social attachment are very strong, but they have yet to be factored in to physical models of crowd behavior.

Inescapable Crushes

When crowds are so dense that movement becomes difficult, social forces become less important. All evidence indicates that it is physical forces that do the damage. We may wish to help others, but we simply can't move in the crush. We may stay calm, or we may panic, but the pressure from surrounding bodies remains the same.

In the disaster at the Who concert, that pressure came from people on the *outside* of the crowd, pressing to get in, unaware of the enormous pressure they were exerting on those on the inside. Video recordings of crowd disasters are now helping researchers to get to the core of the problem, and to understand what can be done about it.

The most thorough study is of the terrible crowd disaster that occurred during the Stoning of the Devil ritual during the Muslim hajj in the city of Mina (just east of Mecca) on January 12, 2006. At least 346 pilgrims died and 286 were injured in a crush that had no apparent cause.

Some three million pilgrims converge on Mecca each year during the hajj. The Stoning of the Devil ritual is the culmination, representing the trials experienced by Abraham while he decided whether to sacrifice his son as demanded by Allah. Pilgrims climb ramps to the multilevel Jamaraat Bridge in the city of Mina and throw pebbles they have collected from the plain of Muzdalifah at one of three large pillars, or *jamaraat*. On the first day, seven pebbles must be thrown at the largest pillar. On the two following days, seven pebbles must be thrown at each of the three pillars (a total of forty-nine pebbles over the three days).

Before the pillars were replaced by substantial walls, it was not uncommon for pilgrims to be injured by pebbles thrown from the other side of the pillar that missed their target. The worst problems, though, have come when people have been trampled by the crowd while trying to approach the pillars.

The tragic events of 2006 were caught on fixed surveillance cameras. Modern image analysis techniques have allowed scientists to scrutinize these recordings and to evaluate the movements of individual pedestrians and small groups of pedestrians. As a result, several formerly unknown features of the behavior of crowds at very high densities have been revealed.

The first is that no matter what the density of people, the crowd as a whole keeps moving. When crowd densities become sufficiently high, however, pedestrians stop intermittently. These stop-and-go waves propagate upstream (against the flow direction) and can be clearly seen when the footage is played faster than real time.

Initially, such stop-and-go waves do not cause severe problems, because people still feel that they have some control over their movements. But when the density corresponds to around one and a half square feet per person, people are moved involuntarily by the crowd. In response, people try to gain space, by pushing other people away, for example. The resulting forces in the crowd add up over large distances. Force chains develop, causing large variations in the size and direction of the force acting on an individual. The video recordings show that, under these circumstances, the crowd mass splits up into clusters of people moving as a block, but relative to nearby clusters. Stress release can be quite unpredictable and uncontrollable, making falling down more likely. The mathematics that describes force chains in crowds is very similar to that used to describe earthquakes, and the consequences of the eventual eruption can be just as severe.

There is very little we can do as individuals when caught in such situations. The best thing to do is to avoid them. Crowd organizers, though, *can* take some action.

To the credit of the Saudi Arabian authorities, action in response to the scientific study of the 2006 hajj disaster was immediate and decisive. A new Jamaraat Bridge with a higher flow and stone-throwing capacity was built, and the design and organization of the plaza around it were modified to balance inflows and outflows. The accumulation of large crowds was prevented. The streets became unidirectional, supporting smooth and efficient flows. Time schedules and a routing plan for pilgrim groups were developed to distribute the flows of pilgrims throughout the day. Moreover, an automated counting system was installed so that crowd densities could be monitored, pilgrim flows could be rerouted, and time schedules could be adapted. Last but not least, an awareness program was implemented to inform pilgrims about the site before and after they arrived.

Overview*

Crowds have emergent complex structures that arise from physical and social forces between individuals. The best way to behave in crowds depends very much on their density. For crowds of low to medium density, the best we can do is be aware of the emergent structures (such as rivers of pedestrians) and use them to our advantage. If we try to do better, we are likely to end up doing worse.

The only exception is when we are in a crowd that is searching for an exit to escape from a dangerous situation. In this case, our best chance, in the absence of additional information, is to go with the crowd 60 percent of the time, using the other 40 percent searching for alternative exits on our own.

When crowds become very dense, we lose much of our control of our own destinies. The best we can do is avoid getting into such crowds. If we are on the outside, we should back off and try to persuade others to do likewise. In this way we can make a small contribution to avoiding catastrophic pressure build-ups in the center of the crowd.

Finally, if there is a warning of danger, act on it promptly; don't wait until you are caught up in the crowd.

---

* Professor Helbing suggests that most of the rules in this summary can be encapsulated by one simple piece of advice, which is equally applicable to walking in pedestrian crowds and driving in crowded traffic: always keep a sufficient distance from others.

FIVE

Group Intelligence: The Majority or the Average?

Our behavior and decisions in crowds are constrained by the fact that we can usually interact only with those nearby. In most other group situations, we can communicate much more freely with all group members to make joint decisions. But how can we use this enhanced communication to make the best group decisions?

There are two basic options: to take a majority vote or to determine some sort of average opinion. Author James Surowiecki has produced a plethora of examples of the latter in *The Wisdom of Crowds.* Recent research has delineated the essential circumstances for this approach to work; more importantly, it has shown *how* it works. In this chapter I examine old and new examples in light of this research and provide an answer to a fundamental question: when should we go with the majority, and when should we take some sort of average opinion?

One of my favorite childhood activities was camping in the Australian bush, the setting for one of my earliest memories of the equal-weight approach. In the dead of night, my father asked me and a group of eight friends to use our compasses to work out where East was. We had only a dim tent light to see by, and some of the compasses were very battered indeed, with sticking needles and loose bearings. Needless to say, our answers were all over the place—they covered a range of some 90 degrees. Dad averaged the lot and drew an arrow in the dirt to show where the average pointed. When the sun came up, the arrow was pointing almost straight at it!

By averaging our results we had achieved an apparently miraculous accuracy. Later that day, on a hike, he asked us whether we thought a wombat or a wallaby (a small kangaroo) was heavier. Six of us thought a wombat was heavier; three of us thought that a wallaby weighed more. This time there was no point in taking an average (what is the average of a wombat and a wallaby—a wannabe?). Instead, the question was simply whether the majority was right or not. As it turned out, the majority was right.

In both cases we had used group intelligence to find an answer. Despite the diversity of our opinions, we had managed to arrive at the right answers. In fact, such diversity lies at the heart of group intelligence. It's a matter of finding the best way to use it—and that depends on the type of problem we are trying to answer.

For problems that involve figuring out the value of something (like a compass bearing or the classic case of the number of jelly beans in a jar) the best way to tackle the problem is to take an average of all the answers. Scientists call these "state estimation" problems. For problems that involve choosing the right answer from among a small number of possible alternatives, majority opinion serves us better. To take the best advantage of either, we need to fulfill just three conditions:

- The people in the group must be willing and able to think for themselves and reach diverse, independent conclusions.

- The question must have a definite answer that can ultimately be checked against reality.
- Everyone in the group must be answering the same question. (This may seem obvious, but it is often possible for people to interpret the same question in very different ways.)

When these three conditions are fulfilled, the mathematics of complexity leads us to three astounding conclusions:

- When answering a state estimation question, the group as a whole will always outperform most of its individual members. Not sometimes. *Always.*
- If most of the group members are moderately well-informed about the facts surrounding a question to which there are several possible answers (but only one correct one), the majority opinion is almost always bound to be right. If each member of a group of one hundred people has a 60 percent chance of getting the right answer, for example, then a rigorous mathematical formulation proves that the answer of the majority has a better than 99 percent chance of being the correct one.
- Even when only a few people in a group are well-informed, this is usually sufficient for the majority opinion to be the right one.

How can we use these principles in our everyday lives?

Taking an Average: The Many Wrongs Principle

Two wrongs may not make a right, but many wrongs can come pretty close. That's the amazing outcome of group intelligence when it comes to state estimation problems.

It's fairly easy to see how it works in principle. In the compass bearing problem of my youth, for example, some of the errors were toward the north, others were toward the south; they largely canceled out when we took an average, so the average was pretty close to the true value.

The pioneering statistician Francis Galton was the first to demonstrate that the bigger the group, the more accurate the guess. Galton was particularly keen on real-life applications of statistics. One famous example concerned the power of prayer. He argued that if the power of prayer is proportional to the number of people praying for a given outcome, then British royalty should live longer than the rest of the population because they were always being prayed for in church. In fact, their lifetimes turned out statistically to be slightly shorter.

Galton, half-cousin of Charles Darwin and member of the upper class, was no democrat, but he was keenly interested in the democratic process, writing that "in these democratic days, any investigation into the trustworthiness and peculiarities of popular judgments is of interest." When an opportunity to use statistics to analyze those judgments came up, he grabbed it literally with both hands.

The opportunity came when he visited the West of England Fat Stock and Poultry Exhibition, held in 1906 in the English coastal town of Plymouth, the town from which Sir Francis Drake famously issued forth in 1588 to defeat the Spanish Armada. Galton was eighty-four years old at the time but as enthusiastic as ever, and fascinated by a guessing competition in which nearly eight hundred people had paid sixpence each for the privilege of guessing the weight of a huge ox after it was slaughtered and dressed.

Galton saw the competition as a mirror of the democratic process because "the average competitor was probably as well fitted for making a just estimate . . . as an average voter is of judging the merits of most political issues on which he votes." He was very interested to discover how the collective guess of the crowd compared to the guesses of its individual members, and after the competition was over he persuaded the judges to lend him the numbered cards on which competitors had written their entries. He returned home triumphantly with an armful of cards that he lined up in order of the weights guessed.

To analyze them, he used the same principle that he had espoused for democratic choice: "one vote, one value." "According to the democratic principle . . . ," he wrote, "the middlemost estimate expresses the *vox populi*, every other estimate being condemned as too low or too high by a majority of voters." The middlemost estimate is what we would now call the "median" (half the guesses are below and half the guesses are above). The guesses for the weight of the ox ranged from 1,074 pounds to 1,293 pounds, and the median was 1,207 pounds; less than 1 percent away from the true weight of 1,198 pounds.*

Galton was astounded by the closeness of the collective estimate to the true value: "This result is, I think, more creditable to the trustworthiness of a democratic judgment than might have been expected." It would be another hundred years, though, before complexity scientist Scott Page came up with the correct mathematical explanation for this phenomenon—one that uses the mean rather than the median.

Page had more than Galton's results to explain. Similar experiments over the years have given very similar results. London architect

* The statistician Karl Pearson recalculated Galton's result in 1924 and reported it as a mean (i.e., average) value, which came out at 1,197 pounds. This is the figure that Surowiecki and others quote in accounts of this incident, but a reading of Galton's original paper shows that, although he was aware that the mean value came even closer to the true value, he still believed that the median was the right one to use because the results were not uniformly distributed. The same argument applies to the assessment of damages by a jury, because one aberrant vote can dramatically affect a mean but it will not much affect the median. Whichever calculation you use, though, it is obvious that the crowd did a lot better than most of the individuals in it.

Matt Deacon, for example, took a glass jar containing 421 pennies to an architectural conference and invited 106 participants to guess the number. The guesses were widely distributed, but their mean value was 419!

Wall Street investment strategist Michael Mauboussin has tested the ability of students at the Columbia Business School to estimate the number of jelly beans in a jar for more than a decade, keeping extensive records. The results from year to year have been remarkably similar. To give an example for one year, in 2007 the number of jelly beans in the jar was 1,116. The average of the guesses was 1,151, and only two of the seventy-three students who participated came closer to the right answer than did the class average.

New York Times columnist Joe Nocera compared the ability of a group to outguess its members to the situation in *Shakespeare in Love* where a series of muddles and mishaps during rehearsals somehow produce a first-class play on opening night. Asked to explain how this happens, the producer in the film explains, "It's a miracle."

The wisdom of crowds is not a miracle. It is simply a matter of statistics. The important thing is that the guesses should be *independent.* If they are not, the crowd rapidly loses its wisdom.

I tested the value of independent guesses on a small scale in my local pub by asking people to guess the number of chocolate-coated pieces of licorice in a small jar, with strict instructions not to let the others know what their guess was. The guesses ranged from 41 to 93, but the average was 60, just 1 away from the actual number of 61. No individual in the group of twenty got this close.

I tried a similar experiment the following week with some mints, encouraging people to discuss their guesses. This time the scatter was much less, between 97 and 112. Unfortunately, there were 147 mints in the jar. Most of the group had been influenced by one strong character, producing an estimate that was quite off.

With independent guesses, the group beats most of its members—not just some of the time, but *all* of the time. This startling truth even applies to weather forecasting. Michigan weather forecaster John Bravender, for example, has pointed out to me in a private e-mail that:

> We have a number of computer models [each designed and believed in by a different forecaster] that may offer a number of different solutions for how the future weather pattern will evolve. Generally, if you average all of them together, you will get the most likely scenario.

Complexity theorist Scott Page, coincidentally from the University of Michigan, has explained *why* diversity of opinion is a key factor in getting the best out of a group in his *diversity prediction theorem.* It relates the collective error of the group as a whole to the average error of its individual members and the diversity of their predictions, or assessments. His theorem simply says:

collective error = average individual error – prediction diversity

The prediction diversity is the spread of the individual guesses. The average individual error is just what it says it is: the average of how far each guess is from the true value. And the collective error is the difference between the average of our individual guesses and the true value.

The calculations are slightly tricky because statisticians use *squared* values of the errors (to get around the fact that some errors are positive and some are negative, depending on whether they are higher or lower than the mean). I give an example in the notes to this chapter to show how the calculation works, but you really don't need to be able to do the calculation to understand the message of the equation.

The message is simple. Just looking at the theorem shows that our collective error as a group *has* to be smaller than our average individual error *because* of the diversity of our answers. The crowd predicts better than most of the people in it. *The group always does better.* Pages puts this neatly: when it comes to determining an average opinion in a state estimation problem, he says that "*being different is as important as being good.*"

The best sort of diversity to have when tackling such problems is cognitive diversity. This encompasses diversity of:

knowledge—specifically, a range of different areas of relevant knowledge within the group
perspectives—different ways of viewing a problem
interpretations—different ways of categorizing a problem or partitioning perspectives
heuristics—different ways of generating solutions to problems
predictive models—different ways of inferring cause and effect

With these in place, it's a matter of taking advantage of the diversity. There is just one caveat, which is to remember that Page's theorem proves only that the group outperforms *most* of its members when it comes to state estimation problems. It does not necessarily outperform *all* of them. If there is an identifiable expert in the group, it may be that they will do better than the group average. If your car breaks down while you are traveling with a mechanic, a poet, and a meteorologist, your best bet will be to consult the mechanic rather than to average the opinions of all three.

That's not to say that experts always do better than the average. There is an increasing body of evidence that suggests that group intelligence often beats them. Firms such as Microsoft, Best Buy, Google, and Eli Lilly have found, for example, that a diverse set of employees with appropriate knowledge can forecast product sales and profits more accurately than so-called budgeting experts can.

A collection of experts can also outperform most, if not all, of the individual experts. Page gives the example of a group of football journalists predicting the top dozen picks in the 2005 NFL draft. Not one of them performed nearly as well as the average of all of them.

Experts are also being replaced by computers when it comes to making rule-based decisions. Computers are now being used in this manner for diagnosing medical and mechanical problems, credit scoring, traffic management, and even literary textual analysis. So what role is left for the poor expert?

According to Mauboussin, experts come into their own in the area between rote rule following and probabilistic prediction—an area in which a combination of knowledge and initiative is required. Mauboussin argues that the best types of experts are those that political and business psychologist Phil Tetlock has identified as "foxes," who have wide knowledge about many aspects of their field, as opposed to "hedgehogs," who have a deeper but narrower knowledge. The claimed advantage is that foxes are able to make more accurate predictions because they have the advantage of diversity built in.

I would modify this with the observation that some hedgehogs are actually like trees with very deep roots that spread underground and pop up in the most unexpected places. These people have a different sort of diversity: the ability to make diverse connections. One of my scientific colleagues, for example, who knows everything there is to know about a particular type of liquid flow called "extensional flow," has contributed not just to the field of physics but also to plastics manufacturing, food production, and even knee surgery—all because his expertise enables him to see the importance of this sort of flow in different contexts. This ability has certainly enhanced his value as an expert.

When we don't have an expert available, we must fall back on the diversity of the group. Taking an average is not the only way to use such diversity. If the problem involves a choice between just a few possible answers, it is majority opinion that gets the nod.

Majority Opinion and the Jury Theorem

The remarkable power of diversity reveals itself fully when it comes to using majority opinion to make decisions. Michael Mauboussin produces a neat demonstration in another experiment with his Columbia Business School students. Each year, just before the Academy Awards are announced, he gets the students to vote on who they think will win in each of twelve categories—not just popular categories like best actor but relatively obscure ones, like best film editing or best art direction. In 2007, the average score for individuals within the group was 5 out of 12. The group as a whole, though, got 11 out of 12 right!

Why is the majority so often right? One reason can be illustrated by the story of the Constitution, and of two of its principle framers, Benjamin Franklin and Thomas Jefferson.

Franklin and Jefferson both spent time in Paris before working on framing the Constitution, which was adopted in 1787. Both of them became involved in discussions with French intellectuals who were primarily responsible for the first French constitution, which was completed in 1789. One of those intellectuals was the Marquis de Condorcet, a corresponding member of the American Philosophical Society, founded by Franklin in 1743 (and still going strong).

Condorcet had begun his career as a mathematician, but when Franklin met him he had been appointed as inspector-general of the Paris Mint at the instigation of the reforming economist Anne-Robert-Jacques Turgot. Turgot didn't last long in the atmosphere of intrigue and double-dealing that characterized Louis XVI's court, but Condorcet prospered. He also became fascinated by the idea that mathematics could be used to support arguments for human rights and moral principles.

Franklin met up with Condorcet many times after he arrived in Paris and was impressed by the progress that Condorcet had made with his "social mathematics," saying at dinners he attended that it "had to be discussed." Nothing was yet on paper, but that soon changed with the publication of Condorcet's remarkable work *Essay on the Application of Analysis to the Probability of Majority Decisions*, published in 1785. There is a copy of the book, signed by Condorcet himself, in Jefferson's library.

Franklin was clearly influenced by Condorcet's ideas, in particular by his mathematical proof of what is now known as "Condorcet's jury theorem." John Adams told Jefferson that Condorcet was a "mathematical charlatan," but this was far from being the case, and Condorcet's theorem is now regarded as a cornerstone for our understanding of democratic decision-making processes.

Condorcet wanted to find a mathematical reason for a rational citizen to accept the authority of the state as expressed through democratic choice. He argued that the best reason would be if his or her individual probability of making a correct choice was less than the collective probability of making a correct choice. His theorem appears to prove that this is nearly always the case.

The theorem in its simplest form says that if each member of a group has a better than 50:50 chance of getting the right answer to a question that has just two possible answers, then the chance of a majority verdict being correct rapidly becomes closer to 100 percent as the size of the group increases. Even if each individual has only a 60 percent chance of being right, the chance of the majority being right goes up to 80 percent for a group of seventeen and to 90 percent for a group of forty-five.

Condorcet's jury theorem looks like a stunning mathematical justification of the power of group intelligence in the democratic process, but it relies on five crucial assumptions, some of which are similar, though not identical, to the elements of cognitive diversity:

- the individuals in the group must be independent, which means that that they mustn't influence each other's opinions

- they must be unbiased
- they must all be trying to answer the same question
- they must be well-informed enough to have a better than 50:50 chance of getting the right answer to the question
- there must *be* a right answer

These requirements mean that the jury theorem is useful only in a very restricted range of circumstances—although it was (and continues to be) a concrete starting point for discussions on how democracy can best be made to work, and on the way that consensus decisions are arrived at in nature. Condorcet even used it after the French Revolution to suggest the best method of jury trial for the king, but his ideas were not taken up in an atmosphere that was more concerned with retribution than with fairness.

Condorcet also invoked the jury theorem in a discussion about the structure of government under the new U.S. Constitution. A point on which all the Framers were firm was that the new government should consist of two houses—a House of Representatives, representing the people, and a Senate, representing the states. When copies of the U.S. Constitution arrived in Paris in November 1787, Condorcet wrote to Franklin, complaining that such a bicameral legislature was a waste of time and money because, according to his mathematical approach to decision making, "increasing the number of legislative bodies could never increase the probability of obtaining true decisions."

The point that Condorcet missed was that the two houses were put in place to answer slightly different questions. The U.S. Supreme Court made this clear in a 1983 judgment about the functions of the two houses when it said, "the Great Compromise [of Article I], under which one House was viewed as representing the people and the other the states, allayed the fears of both the large and the small states." In other words, the House of Representatives is there to ask,

"Is X good for the people?" while the Senate's job is to ask, "Is X best implemented by the federal government or by the states?" The fact that the two houses are answering slightly different questions negates Condorcet's argument that one of the houses is redundant.

It might appear that the jury theorem is more relevant to the functioning of juries themselves, but here again it is a matter of how juries are set up. To take maximum advantage of group intelligence, jurors need to be truly independent, which means that each would need to listen to the arguments of both sides and then make a decision without discussing it with the other jurors. The decisions would then be pooled, and the majority decision accepted.

Condorcet suggested that Louis XVI's jury be set up in this way, but his ideas were rejected, and as far as I can find there have been no tests of his proposal since, in France or elsewhere. It does seem a pity, because discussions between jury members before coming to a decision mean that one of the main foundations of group intelligence (that of independence) is lost. Discussions certainly have their value—allowing people to change their minds under the influence of reasoned argument—but other forces can also be at work. One of these is the social pressure to conform with other members of the group that goes under the name of "groupthink," and which I discuss in the next chapter. So long as members of juries continue to thrash out the merits of a case between themselves before coming to a conclusion in the manner depicted in the film *Twelve Angry Men*, the jury theorem will largely be irrelevant to their deliberations.

It comes into its own, however, when applied to the game show *Who Wants to Be a Millionaire?* although it turns out that our collective judgment is even more reliable than the theorem suggests. James Surowiecki points out that the "Ask the Audience" option consistently outperforms the "Call an Expert" option. This group of "folks with nothing better to do on a weekday afternoon" produces the correct answer 90 percent of the time, while preselected experts can only manage 66 percent.

It seems like an ideal case for the jury theorem. The selections are independent. The audience is presumably unbiased. Its members are all trying to answer the same question, and the question has a definite right answer.

The assumption that all members of the audience need to have a better than 50 percent chance of getting the answer right, however, is *not* necessary. Close examination reveals that their group intelligence still works even if only a few people know the answer and the rest are guessing to various degrees.

To see how this works, try the following question, originated by Scott Page, on your friends. Out of Peter Tork, Davy Jones, Roger Noll, and Michael Nesmith, which one was not a member of the Monkees in the 1960s?

If you ask this question of 100 people, one possible scenario is that more than two-thirds (68, say) of them will have no clue, 15 will know the name of one of the Monkees, 10 will be able to pick two of them, and only 7 know all three. The non-Monkee is Roger Noll, a Stanford economist. How many votes will he get?

Seventeen of the 68 will choose Noll as a random choice. Five of the 15 will select him as one choice out of three. Five of the 10 will select him as one choice out of two. And all of the 7 will choose him. This gives a total of 34 votes for Noll, compared to 22 for each of the others—a very clear majority.

So group intelligence can work in this case with only a few moderately knowledgeable people in the group. It would even have a fair chance of working if 68 people had no clue and the remaining 32 only knew the name of one Monkee. One-third of these (11 to the nearest whole figure) would choose Noll as the exception, giving an average total of 28 votes for Noll and 24 for each of the others.

Statistical scatter makes this prognostication less sure, but with increasing group size the difference becomes more meaningful.

When it reaches the millions, the majority vote can provide a very sure guide, which is why search engines such as Google, Yahoo, and Digg.com use it as an important guide in their ranking algorithms.

The jury theorem is fine for the very restricted conditions under which it applies, and it is a necessary starting point for thinking about majority voting in different contexts. Modern analysis has built on it and shown that group intelligence can be even more powerful than the theorem suggests—so long as we get the conditions right. As I show in the next chapter, this is particularly important when it comes to getting groups to reach a consensus.

SIX

Consensus: A Foolish Consistency?

When I was young and idealistic (I am now old and idealistic) I was involved in the formation of a new political party in Australia. Eager to do my bit, I volunteered to be its policy coordinator.

The other parties didn't have policy coordinators, but we were determined to be as democratic as possible, and that meant taking account of the views of all of the members. Luckily there weren't very many members, but even so it was an onerous task. I decided to make it easier and more efficient by using the Delphi technique, a tool used in business to help a group progress from consultation to consensus.

The basis of the method is working the consultation in a series of steps:

1. Circulate the problem to group members.
2. Collate the responses, suggestions, and supporting arguments.
3. Send the collation back to the group members and ask them to rate the suggestions.

4. Repeat steps 2 and 3 if necessary until a consensus is reached.

I kept it simple. First I sent out questionnaires requesting suggestions for a policy on some issue. A group of us would then summarize the answers, prepare a set of options that reflected the range of opinions, and send out the options for the members to vote on.

Our summaries, though, were treated with suspicion. One member even wrote that they should be compared with reports of parliamentary debates that were written by the great lexicographer Samuel Johnson in the eighteenth century. This was not as flattering as it sounded, because Johnson captured the essence of what was said in the debates between the Whigs and the Tories, but he invented the actual dialogue, claiming that he "took care that the Whig dogs did not get the best of it."

Our members were worried that we were doing something similar, writing the summaries with a bias toward our own views. We tried to assure them that we weren't and that they were welcome to examine the voluminous pile of correspondence from which we had worked. Needless to say, that didn't cut much ice, and the party soon collapsed under the weight of its own attempts at democracy.

Our dilemma reflected one of the biggest problems facing any group: how to make the transition from diversity-based problem solving to a consensus agreement on a course of action, which inevitably reduces diversity. Whether we take an average, accept the majority opinion, or seek out the most knowledgeable person and accept their guidance, *someone* has to sacrifice their opinion so that the whole group can benefit from its group intelligence.

There are several ways to make the transition from diversity to consensus. Here I focus on three. One is to follow the example of nature and go along with what the majority of our neighbors is doing. Another is to debate the issues and come to a reasoned agreement. The third is to use swarm intelligence.

Quorum Responses: The Example of Nature

One way to achieve consensus is to follow the example of others who appear to know what they are doing. When we select a well-worn track over a less-used one on a hiking trail, for example, we are presuming that the well-worn one is the more likely to have been made by people who knew where they were going. Similarly, when ants use pheromone concentration as the criterion for choosing between routes to a food source, they use the decisions of ants that have previously used the trails to guide their own choice.

Our main aim is to avoid getting lost. Ants and other animals that travel in groups have more serious purposes—to find the best food sources, to obtain the best shelter, and above all to avoid getting eaten during the search.

These animals can improve their chances by copying the actions of better-informed neighbors. But how are they to know which neighbors are better informed? Their only real clue lies in how many others are also copying them.

It's a clue that the people of Moscow used during the 1980s when basic goods were in perennial short supply after the collapse of communism. If you were walking down the street and saw one or two people standing around outside a shop, you might have walked on. But three or four people was a signal that the shop had something available to sell, and others would hasten to join the line in a cascade effect that rapidly produced a longer line, although hardly anyone in the line knew what was for sale!

This cascade effect is known to animal behaviorists as a *quorum response*. Put simply, the group arrives at a consensus in which each individual's likelihood of choosing an option increases steeply (nonlinearly) with the number of near neighbors already committed to that option (the neurons in the human brain show a similar sort of response to the activity of neighboring neurons).

The quorum response is an example of complexity science at work. All of the basics are there—local interactions, nonlinear responses, positive feedback, and the emergence of an overall pattern. It's just a matter of whether the pattern is a useful one or not.

In nature, the quorum response can lead to very useful patterns. One example is when it ensures the cohesion of the group, which has advantages such as protection from predation and consensus in group decisions. The sharp nonlinearity of the response also ensures faster decision making and (if the individuals initiating the response are well informed) a higher rate of accuracy of the group's decisions.

Animals can use the quorum response to make a choice between speed and accuracy. Speed is preferable if time is of the essence. Computer modeling has shown that the requisite speed can be achieved by responding to the actions of a smaller number of neighbors. If there is plenty of time available, the reverse effect can be achieved by waiting until more neighbors have committed to a choice before making the same choice.

Many of us use the quorum response when we are choosing a place to eat during a long drive. If there are no cars parked outside a roadside diner, we are likely to drive by. If there are one or two cars, we may pull in. If there are plenty of cars we are almost certain to pull in, on the basis that this must indicate that it is a good place to eat, or perhaps that there are no more places farther down the road. Wherever we pull in, we have added another car to the total, thus increasing the chances that others will pull in as well.

We can also use the quorum response to make a trade-off between speed and accuracy. If I am feeling particularly hungry, for example, I am likely to take a risk on quality and pull in at a restaurant that has just a couple of cars outside. If I am less hungry, I am more likely to drive on, looking for one with more cars outside.

I am trusting that the car owners will have used some form of independently gathered information to choose this particular restaurant. They may have been there before, or read a review, or been told about it by their friends. If they were all using the same strategy that I was—relying on the presence of other diners as an indication of quality—we could all be in for a shock. In these circumstances, the first few would have chosen the restaurant at random, and the rest of us would have followed them blindly, like the lemmings filmed while cascading over a cliff in the film *White Wilderness.*

This phenomenon of interdependence without individuals having reliable, independently gathered information is called, rather oddly, an *informational cascade.* (Maybe a better term would be *disinformational cascade.*) Some unscrupulous owners have been known to initiate such a cascade by parking a group of their own cars outside their restaurants just to encourage people to pull in. This points up a basic problem with quorum responses: for them to result in the desired outcome, we must be able to trust the truthfulness and knowledge of those whose decisions we are choosing to copy.

Sometimes these qualities are obvious. If we are looking for shelter from a sudden downpour in an unfamiliar environment, for example, it's usually a good idea to follow the crowd, which is likely to be made up of locals who know what they are doing.

In general, it's a good idea to follow the crowd if its purpose is the same as ours, whether that is looking for food, shelter, or a bargain. Animals such as cockroaches, ants, and spiders don't do a lot of shopping, but they use the quorum response very successfully in their search for food and shelter. We can do the same, and there are many circumstances in which we can do a lot better.

The quorum response has its uses, but we can often improve on it by factoring in our own judgment. It's the interplay of independence and interdependence that gives us our best chance. When we check out how many people are using a restaurant, for example, we can supplement that information by looking into the kitchen to see if it is clean, or just taking a look at the customers' faces and the appearance of the food on their plates.

We can supplement it further by asking someone who has just come out of the restaurant what they thought of it. They may not tell us the truth, of course, especially if they are friends of the manager. That would not matter if my wife, Wendy, were with us, because she seems to be able to detect a lie instantly through body language. She can certainly see through *me* all right. She may even be one of the 0.25 percent of people that scientists have found can do it almost *all* the time.

Regardless of how we gather our information and form our opinions, when we are part of a group we must translate everyone's information and opinions into some sort of consensus for action.

If the group is under the control of a dictator (benevolent or not), achieving consensus presents no problem. When I was a child and my family went out to dinner, we would squabble endlessly about what sort of restaurant we wanted to go to until my father threw up his hands in frustration and said, "*This* is where we are going!" At other times, though, once we had all expressed our opinions and advanced our arguments, he would put it to a vote.

Putting it to a vote is the democratic way of reaching a consensus. There are many hidden traps, however, both in choosing an appropriate voting method and in avoiding groupthink (see the section later in this chapter). Here I examine the problems with voting methods, and what (if anything) we can do about the dreaded groupthink.

Voting Methods

Evidence shows that all voting methods are flawed, so we may as well choose something simple to suit the particular situation and get on with it.

The idea of getting together to vote on an issue goes back two-and-a-half thousand years, to a time when the city of Athens in ancient Greece was laying the foundations for Western civilization. The citizens of Athens had two great ideas about selecting their politicians. The first was to choose them by lottery from whoever put their name forward. The second was to get rid of the worst ones by an annual process of negative voting.

Negative voting consisted of writing the name of the disfavored politician on a broken piece of pottery called an *ostrakon.* On the appointed day the citizens who wished to vote came to the civic center of Athens and handed in their ostrakons to be counted. So long as a quorum of 6,000 votes was cast (out of the 50,000 or so citizens who had voting rights), the politician who had the misfortune to get the most votes was barred from the city for ten years—in other words, they were *ostracized.*

The main idea of the system was to break voting deadlocks that barred the way to consensus decisions. But it all fell apart during the wars with the city of Sparta. An Athenian politician called Nicias had brokered a fragile peace, but another one called Alcibiades wanted to resume all-out war. The population was evenly divided on the issue, and a vote was called to ostracize one or the other and open the road to a decision.

Alcibiades and Nicias responded by each urging their supporters to ostracize a third politician, called Hyperbolus. Hyperbolus was exiled, and the issue remained unresolved. After this disastrous result, the population saw that ostracism could be manipulated, and although it remained in the statute books, it was never used again.

Manipulation, though, is only one problem that voting systems face. Another is mathematics, in the form of the *voting paradox.*

The paradox was discovered by the Marquis de Condorcet, who had more fun with voting systems than most politicians seem to have. Condorcet noticed that majority voting can lead to a paradoxical outcome when it comes to choosing between three or more alternatives.

Let's call the alternatives A, B, and C. What Condorcet proved was that even though each voter has a definite order of preference, when all of the votes are put together it is still perfectly possible for A to beat B, and B to beat C, but C to beat A!

The voting paradox is not the only one that arises when it comes to choosing between three or more alternatives. If majority voting is used, so that the winning alternative is the one that gains the most votes, it is not only possible but normal for *the winner to be the choice of the minority*, as experience in many elections has shown.

Small children can be quicker than adults in catching on to these paradoxes and their consequences, as Illinois mathematician Donald Saari found when he presented them to a class of fourth graders.

Saari used the example of a group of fifteen children who have to vote on which television show to watch in the evening (they are allowed only one show). He asked the fourth-graders what show they should watch if the voting went like this:

Number of Children	First Choice	Second Choice	Third Choice
6	*ALF*	*The Flash*	*The Cosby Show*
5	*The Cosby Show*	*The Flash*	*ALF*
4	*The Flash*	*The Cosby Show*	*ALF*

Majority voting suggests that *ALF* should get the nod, but the fourth graders vigorously disagreed. "*Flash*!" they cried—and they had a point. *The Flash* might come last in the majority stakes, but nine of the fifteen prefer it to *ALF*, and ten of the fifteen prefer it to *The Cosby Show*.

This simple story points up the problem with majority voting—a majority of the voters can end up with the candidate that they *don't* want. Condorcet's voting paradox, though, points up an even more perplexing problem.

Suppose the children's voting had gone like this:

Number of Children	First Choice	Second Choice	Third Choice
5	*ALF*	*The Cosby Show*	*The Flash*
5	*The Cosby Show*	*The Flash*	*ALF*
5	*The Flash*	*ALF*	*The Cosby Show*

Now *ALF* is preferred over *The Cosby Show* by 10 votes to 5, and *The Cosby Show* is preferred over *The Flash* by 10 votes to 5. So *ALF* should be preferred over *The Flash*, right? Wrong! As the fourth graders noticed (and pointed out vociferously), a simple count shows that *The Flash* is preferred over *ALF* by 10 votes to 5!

The paradoxes that the fourth graders picked up on are not just academic puzzles. They often arise in the real world of voting, whether the vote is for electing politicians or for making a decision in a committee or other group. But it gets worse. Much worse.

An unwelcome further complexity was discovered by the Nobel Prize–winning economist Kenneth Arrow, a founding father of the Santa Fe Institute. Arrow showed in 1950 that Condorcet's paradox is not just an exception to the rule; it is part of a wider set of paradoxes that *are* the rule.

Arrow first looked at what we might want from an ideal voting system. His full list of criteria (paraphrased here from his more technical descriptions) was:

1. *Completeness*: If there are two alternatives, the voting system should always let us choose one in preference to the other.
2. *Unanimity*: If every individual prefers one alternative to another, then their aggregated votes should reflect this choice.
3. *Non-Dictatorship*: Societal preferences cannot be based on the preferences of only one person regardless of the preferences of others.

4. *Transitivity*: If the aggregated votes show that society prefers choice X to choice Y, and choice Y to choice Z, then they should also produce a preference for choice X over choice Z.
5. *Independence of Irrelevant Alternatives*: If there are three alternatives, then the ranking order of any two of them should not be affected by the position in the order of the third.
6. *Universality*: Any possible individual ranking of alternatives is permissible.

Some of these criteria may look trivial. All of them look eminently reasonable for voting in a democracy. Yet Arrow proved in his impossibility theorem (otherwise known as "the paradox of social choice") that we can't have them all at the same time. If we have majority voting, for example, then Condorcet's paradox shows that we can't have transitivity. If we use my father's dictatorial approach, then criterion 3 goes out the window.

There really is no way out of it. However we wriggle, and no matter what voting system we adopt, one of Arrow's criteria has to be abandoned.

In his 1972 Nobel memorial lecture, Arrow said that the "implications of the paradox of social choice are still not clear. Certainly, there is no simple way out. I hope that others will take this paradox as a challenge rather than as a discouraging barrier."

Even before the discovery of the paradox in 1950, it was perfectly clear that democracy must involve compromise. Former British Prime Minister Winston Churchill put it best in a 1947 speech when he said, "Many forms of Government have been tried, and will be tried in this world of sin and woe. No one pretends that democracy is perfect or all-wise. Indeed, it has been said that democracy is the worst form of government except all those other forms that have been tried from time to time."

Arrow's proof uncovered one of the difficulties for democracy. His discovery has formed the bedrock for discussions about the best compromises for implementing democracy ever since. From the point of view of small-scale practical democracy, the paradox means that we can never hope to achieve perfection, and the best we can do is choose a simple voting system that seems reasonable for the purpose, and stick with that. My father, at first, used a family majority vote to choose a restaurant. Later, he let each of us be a dictator in turn. Neither system was perfect, and each appeared to work equally well.

When it comes to larger-scale democracy, my personal opinion is that simplicity is overrated. First past the post, for example, is obviously simpler than a preferential voting system, in which candidates are ranked in order of preference and the least favored candidate is removed from the bottom of the list and the second preferences of the people who voted for him are distributed among the other candidates, with the process being repeated until a clear winner emerges.

Few would disagree that a preferential system is fairer. First-past-the-post methods suit major parties better, though, by rendering votes for the minor parties ineffective, so it is likely to continue in countries like the United States and the United Kingdom.

Even with the most perfect voting system in the world, there is still the human dimension to consider. People may vote tactically, form voting blocs, or be influenced by personalities rather than issues. All of these are the province of game theory, which is the subject of my book *Rock, Paper, Scissors.* There is one human issue above all, though, that consistently undermines our efforts to use group diversity to come to the best consensus. That issue is groupthink.

Groupthink

Groupthink is the phenomenon where social pressures within the group push its members into "a pattern of thought that is characterized by self-deception, forced manufacture of consent, and conformity to group values and ethics." Members of a group are drawn into agreeing to a common position and sticking to it through thick and thin. The outcome can even be MAD (mutually assured delusion), which occurs when group members deny evidence that those outside the group can plainly see, stick to beliefs that have little or no basis in fact, and fall "prey to a collective form of overconfidence and willful blindness."

One often-quoted example is the delusion among those close to George W. Bush during his presidency that an invasion of Iraq (in March 2003) would be a short affair because American troops would be welcomed as liberators by a grateful population. According to investigative reporter Bob Woodward, the factors that encouraged groupthink on this occasion included "the support, the encouragement of [Vice President Dick] Cheney, the intelligence community saying Saddam [Hussein] has weapons of mass destruction . . . and [President George W.] Bush looked at this as an opportunity [to fulfill a private dream]."

Other examples include the $50 billion scam Bernard Madoff operated, in which investors collectively deluded themselves into thinking that he must be cheating on their behalf rather than his own, and the activities of loan institutions that led to the credit crunch, in which they collectively convinced themselves that house prices would keep rising without end, so that toxic loans would ultimately lose their toxicity because of a continuously rising market.

When Yale psychologist Irving Janis coined the term *groupthink* in 1972, he listed its main characteristics as

- *Pressures for uniformity*, such as the threat or actual application of sanctions that makes people feel excluded if they disagree with its way of thinking and its conclusions.
- *Closed-mindedness within the group*, so that any doubt is rationalized away.
- *An overestimation of the group* as strong, smart, morally superior to other groups, or even invulnerable.

The Nobel Prize–winning physicist Richard Feynman experienced all of these when he joined the committee investigating the 1986 *Challenger* disaster. The committee chairman, former secretary of state William Rogers, commented that "Feynman is getting to be a real pain" after Feynman decided to conduct his own investigations rather than simply sit in committee meetings.

Feynman's investigations consisted in talking to the scientists and engineers who had actually worked on the project. He soon found that there were very diverse opinions within NASA about the safety of the shuttle. In his contribution to the final report he said: "It appears that there are enormous differences of opinion as to the probability of a failure with loss of vehicle and human life. The estimates range from roughly 1 in 100 to 1 in 100,000. The higher figures come from working engineers, and the very low figures from the management."

If the management had been aware of this diversity of opinion and used it to make better decisions, it is possible that the disaster would never have happened. The management, however, was particularly subject to the second characteristic of groupthink, closed-mindedness within the group. This led to a situation identified in the final report: although both NASA and its contractor, Morton Thiokol, knew that there was a design flaw, they chose to ignore its potential for disaster, which reflects the third characteristic of groupthink: an overestimation of the group's ability.

The true basis of the failure came to light only because Feynman acted independently by talking to the scientists and engineers. The credit does not all go to Feynman, though. He recognized later that the scientists and engineers with whom he had talked had *led* him to his conclusions. His presence and questions gave them the confidence to escape from the stultifying atmosphere and pressure of groupthink within NASA, at least to the extent of pointing Feynman in the right direction.

Feynman might never have come to his conclusions if he had not evaded the iron fist imposed by Rogers on his commission members. Rogers appears to have decided early on to focus on the NASA administration and ignore the technical details. When he found out that Feynman had been talking to the technical people, he ordered him to stop. By that time, though, Feynman had all the information he wanted.

It all came to a head at a televised meeting of the commission. Feynman produced a sample of the material that was supposed to provide a flexible O-ring seal for the fuel tank from which leakage had precipitated the catastrophic explosion. He then compressed it with a clamp and dunked it in the glass of ice water that had been provided for him to drink and said: "I took this stuff that I got out of your seal and I put it in ice water, and I discovered that when you put some pressure on it for a while and then undo it, it does not stretch back. It stays the same dimension. In other words, for a few seconds at least and more seconds than that, there is no resilience in this particular material when it is at a temperature of 32 degrees."

In other words, a material that was supposed to provide a flexible seal became hard and brittle when it was just dunked in ice water, let alone exposed to the temperatures of deep space, where it would be expected to become even more brittle.

Feynman's actions appeared to demolish the groupthink that was driving the commission, which had been focused only on administrative issues such as poor communication and underlying procedures. The difficulty of truly annihilating groupthink, though, was demonstrated by the fact that the commission's final report *still* focused on these issues. Only after Feynman threatened to remove his name from the report was he permitted to add an appendix that addressed the technical issues.

The appendix had a belated effect when the U.S. House Committee on Science and Technology reviewed the Rogers report and concluded that "the underlying problem which led to the *Challenger* accident was not poor communication or underlying procedures as implied by the Rogers Commission conclusion. Rather, the fundamental problem was poor technical decision making over a period of several years by top NASA and contractor personnel, who failed to act decisively to solve the increasingly serious anomalies in the Solid Rocket Booster joints."

Subsequent events demonstrated the insidious hold that groupthink exerts further. Despite the damning House committee report, the groupthink inherent in the NASA management culture continued. This time the technical issue was the heat-insulating foam on the shuttle's tanks. Pieces kept coming off. According to NASA's regulations, the issue needed to be resolved before a launch was cleared, but launches were often given the go-ahead anyway.

The sickeningly inevitable result was the space shuttle *Columbia* disaster of February 1, 2003, when a piece of foam came off during the launch and struck the leading edge of the left wing, damaging the wing's thermal protection system. As a result, during reentry the wing overheated and the shuttle disintegrated.

The subsequent accident investigation board was again highly critical of NASA's decision-making and risk-assessment processes. It concluded that the organization's structure and processes were sufficiently flawed that compromise of safety was to be expected no matter who was in the key decision-making positions. Not much seemed to have changed in the culture at NASA, least of all the overconfidence that stems from groupthink.

Feynman's example shows, however, that although it is difficult to escape from groupthink, it is not always impossible. The key elements (apart from an iron will) are (1) getting out of the group environment for a while, (2) doing some independent thinking, and (3) committing oneself to the conclusions of that thinking before returning to the group to share your conclusions.

That's not to say that the effects are always what one wants them to be. Feynman wrote a number of short scientific reports during his investigations, sending them to the commission secretary for distribution. When he asked what his fellow commission members thought of them, he found that they had simply been filed.

I did something similar, and with even less success, when I was invited to join a panel of scientists organized by the U.K. Office of Science and Technology to "brainstorm" possible future growth areas in research. Our deliberations, and those of similar panels, were to be taken quite seriously, with money at stake in the form of investment in particular areas.

Even though I don't believe that the path of future research can be foretold in this way (or any other way), I was flattered by the invitation and duly joined the group. Not only that, I soon found myself enthusiastically joining in with the group ethos to produce ideas about where significant research advances were likely to come from. The effect of groupthink was so strong that most of us were convinced that our performance was far superior to that of other groups. Only when I came away from the group after it had met did I collect myself and start to think how silly it all was. I immediately sat down and sent an e-mail to this effect to the organizers and my fellow group members, but the only obvious outcome was that I was never invited back.

Groupthink might still beat me, however, in the form of a belief that permeates government and the community, and which is now starting to pervade science itself. This is the belief that useful science can come only from focusing on its potential applications right from the start. If we had always worked from that belief, and not also asked penetrating questions about how the physical and biological worlds work, we would not have X-ray machines, antibiotics, radio, television, or cars with inflatable tires, to name just a few inventions that stemmed from questions that never had these ends in view.

Janis argued that groupthink was mainly the province of "very high echelon groups," such as those that determine government policy, in which the perks of membership are at "intoxicating levels." He also believed that groupthink was mainly confined to circumstances in which (1) the group insulates itself from outside criticism, (2) there is high stress from a perceived external threat, and (3) there is very strong group cohesion.

But Janis was wrong. Groupthink is everywhere, and it is especially virulent in its ability to affect our attitudes toward each other.

A demonstration of how this can happen emerged from a study conducted by psychologists Donald Taylor, from McGill University, and Vaishna Yaggi, from the University of Delhi. They asked volunteer Hindu and Muslim students in southern India to read stories that showed members of each group in either a good or a bad light. They then asked the students what they thought might be the cause of the good or bad behavior of the characters in the story.

If the story was about bad behavior by a member of their own social group, the failings of the characters were conveniently attributed to external causes. If it was about bad behavior by a member of the other group, the failings were seen as typical of that group. When the stories were about good behavior, the reverse was true. Students saw good behavior as a characteristic of their own group, but if a member of the other group behaved well, it was due to some external factor having nothing to do with the person themselves!

Such attitudes can reflect simple pride in the traditions and attitudes of a society, or they can take us down the terrible paths of racism, strident nationalism, and religious bigotry. They are part of our complex cultures, but they can make it difficult to achieve the maximum potential of that culture, because the group attitudes they represent are the antithesis of the diversity that gives us our best chance of making good group decisions.

Unfortunately, societies are infested with groupthink. It exerts its menacing influence in families, local communities, and gangs. It can also make for trouble at parties, such as one that was organized over the Internet by teenager Corey Delaney, which resulted in the near destruction of his parents' house by drunken youths. The social pressures in this case drove the teens (mostly testosterone-filled boys) to push each other to more and more senseless acts of vandalism, all in the name of fun. One curious and rather repugnant outcome of this episode was that a publicity agent sought Delaney out and set him up as a professional organizer of parties!

What can we do to avoid such damaging effects of groupthink? It is a difficult question, because group decision making must always involve conditioning our choices on those of others. Yet the moment we do this we lose some of our independence, and the group loses some of its diversity. How can collective decisions preserve independence but still come to a final consensus? As pointed out by consensus specialist David Sumpter, this is the paradox that lies at the heart of groupthink, as it does in all group decision-making processes.

Can the paradox be overcome? Irving Janis thought so, and he argued for a process reminiscent of the Delphi technique. His idea was simple: get the individuals in a group to collect information independently (as Feynman did) and work out a recommended course of action, then present that course of action and the reasons for it to a smaller number of centralized evaluators.

It sounds ideal in theory, but would *you* trust a central evaluator to truly represent your views? The members of my political party certainly didn't. Neither did Feynman after he found that his scientific reports had been filed away instead of being distributed. Both cases illustrate a problem that has been understood by game theorists for years: an independent arbiter can sometimes help to resolve a problem, but in many cases the arbiter becomes part of the problem instead.

If we can't rely on leaders or independent arbiters to take proper account of our diversity, what else can we do to smooth the path from diversity to consensus? One option is to use swarm intelligence.

The Business of Swarm Intelligence

Swarm intelligence is subtly different from group intelligence. The latter, as we've seen, is an approach to problem solving that takes advantage of the diversity within a group. Swarm intelligence is a spontaneous phenomenon that emerges from local interactions between individuals in a group. Swarm businesses that take advantage of both are now emerging.

The archetypal example is the Web. According to researchers Peter Gloor and Scott Cooper from the MIT Sloan School of Management, who specialize in the use of collective intelligence in business, the explosive expansion of the Web in its early days arose for the following reason:

> If someone wanted to pursue a useful idea for extending the project—for instance, to include Web browsers and servers—the swarm embraced and supported the effort. There was no managerial hierarchy or proprietary ownership of ideas. Everyone cared deeply about the cause, not about rank, salary, status or money. They just wanted to get the job done, and in the end they changed the world with their innovation.

Diversity solved the problems and provided the innovations. The principles of complexity science did the rest, with a wonderfully intricate network emerging spontaneously from local rules of behavior and interaction.

Traditional businesses are following suit. BMW, for example, posts engineering challenges on its website, so customers and company designers collaborate to arrive at innovative solutions. Businesses such as Ford, Boeing, Procter & Gamble, Beiersdorf, and Chevron-Texaco have taken advantage of advances in complexity science to design new swarm business approaches for some of their operations. The list of traditional companies that are taking up this innovative approach continues to grow.

Cooperatives are another example of the combined effects of group intelligence and swarm intelligence. These are businesses that are jointly owned and democratically controlled by their members. According to the website of the worldwide cooperative movement,

> Co-operatives are trading enterprises, providing goods and services and generating profits, but these profits are not taken by outside shareholders as with many investor owned businesses—they are under the control of the members, who decide democratically how the profits should be used. Co-operatives use their profits [both in industrial and developing countries] for investing in the business, in social purposes, in the education of members, in the sustainable development of the community or the environment, or for the welfare of the wider community.

Today the cooperative movement is "a global force and employs approximately 1 billion people across the world. The UN estimates that the livelihoods of half the world's population are made secure by co-operative enterprise."

In Switzerland, the two million customers of the giant Migros supermarket chain are members of its cooperative enterprise (out of a total population of seven million!) and have helped it develop through a process of self-organization.

In all of these cases the individuals who make up the swarm see themselves as *stakeholders* rather than *shareholders*. A stakeholder is defined as any party that can affect or is affected by the innovation.

In a traditional business, this includes not only shareholders but also employees, customers, suppliers, partners, and even competitors.

Changing the view changes the balance. Shareholders see themselves as outsiders. Stakeholders see themselves as insiders. A shareholder will sell shares or agree to a takeover if she sees that there is a profit to be made. A stakeholder will be less willing to do either, because he would thereby be changing something or losing something of which he is a part.

Stakeholders see themselves as members of a swarm, which may be akin to a sense of family. Whatever the reason for people wanting to belong to a swarm, however, businesses that take advantage of its power are taking off. Some of these, such as Wikipedia and Project Gutenberg (a venture to make classic literature available on the Web) are not-for-profit entities. Others, such as Amazon.com, aim to make a profit by using swarm principles that include allowing people to post product reviews for other customers to read and making suggestions about what customers might like that are based on what individuals with similar buying patterns have bought.

Swarm ideas are not confined to single businesses. They can equally be applied to groups of businesses, as is the case with Amazon.com. The main prerequisite is that favorable conditions should be set up to provide a platform on which those businesses can interact in a swarmlike manner.

According to Gloor and Cooper, on whose ideas the above discussion is based, the principles of a swarm business are radically different from those of a traditional business in three important ways:

- *Swarm businesses gain power by giving it away.* In other words, the power resides with the stakeholders, not with the business itself. This is how businesses like Amazon and eBay make their money. Both offer a business platform—that is, an environment in which more traditional retailers can develop a market. They provide the facilities for the establishment of a network and then let buyers and sellers get on with it while facilitating functions like product representation, regulatory compliance, risk management, and conflict resolution.

- *Swarm businesses are willing to share with and support the swarm.* One example is open source software. IBM, for example, spends $100 million per year to support the development of the freely available Linux operating system. The long-term outcome is that IBM gets more sales for its products that use Linux.
- *Swarm businesses put the welfare of members of the swarm ahead of making money.* For example, the giant life sciences company Novartis AG, which was created by a merger between Sandoz and Ciba-Geigy, set up a venture fund in 1996 to encourage researchers and engineers who had lost their jobs as a result of the merger to start their own companies. The resultant swarm of companies ultimately paid real dividends back to the fund, but this wouldn't have happened without the philanthropically based gamble in the first place.

Could we use the principles of a swarm business in our personal lives? One way would be to make more effective use of the various networks to which we all belong, from family to Facebook, and from close friends to our wider circles of acquaintances and contacts. In the next chapter I examine how such networks operate as complex adaptive systems, and how we can use their guiding principles most effectively.

SEVEN

Crazes, Contagion, and Communication: The Science of Networks

When I was barely into my teens, my father took me aside for a man-to-man talk. Embarrassed, I followed him into his study, expecting a talk about the birds and the bees, a subject, I was convinced, that I already knew more about than he did. But he wanted to talk about my future.

His talk was full of good advice, some of which I have followed. The piece of advice that I remember most was that the secret of success in life is to have good connections. What he was trying to tell me was that if you don't know someone who can help with a problem, it's worth knowing someone who might know someone else who could help.

He built our first house this way, doing the bulk of the work himself but able to call on friends and friends of friends in the network he had built up over the years. The network was a strong one because the help was two-way. Everyone knew that he could repair their watches and jewelry; they in turn were willing to offer their services as carpenters, plumbers, and electricians.

It wasn't a particularly big network as networks go—hardly a hundred people, most of them living within a couple of miles of each other, most of them knowing each other. The network of scientific friends and collaborators that I have built up during my career isn't much bigger, and most of us also know each other. If I have a math problem that I can't solve on my own (this applies to most math problems) I know where to go for help. If someone wants to use one of the instruments I have developed, they in turn know where to come.

The fact that many of us are in different countries doesn't matter. We are still a small, tightly interconnected network. We are also connected to a number of larger networks. The World Wide Web, and the Internet of which it is a part, are the biggest, but we also use telephone networks, delivery networks, airline networks, and road networks to pass information and material goods around.

Such networks are self-organized to various degrees—"a web without a spider" to use network science pioneer Albert-László Barabási's apt description. Some, like the Internet and the Web, seem to have taken on a life of their own. All have properties that belong to the network as a whole rather than to its individual members, such as the famous six degrees of separation by which everyone in the world is supposed to be connected. It is only since the late 1990s that such properties, and their relevance to our everyday lives, have begun to be properly understood.

What Is a Network?

Samuel Johnson defined *network* in his famous dictionary as "any thing reticulated or decussated, at equal distances, with interstices between the intersections." Johnson's definition of a network reflects a genuine struggle with a complex concept; in fact, one that lies at the heart of complexity.

If you cut out the bit about equal distances, it doesn't look like a bad definition. According to the *Oxford English Dictionary*, *decussated* just means "formed with crossing lines." *Reticulated* means "constructed or arranged like a net." And *interstices* are gaps or holes.

Johnson's definition, though, misses the real point of a network. It's not the holes that matter; it's the links between the points of intersection that make the whole thing work. If it's a road network, the links are the roads themselves, and the intersections are where they cross. If it's an airline network, the links are the airline routes, and the intersections (technically called *nodes*) are airports.

If it's people, then the people themselves are the nodes. If they know each other personally, we say that there is a link between them. Sociologists draw this set of connections as a *sociogram*, which permits immediate visualization of personal relationships—who is linked with who, whether someone is socially isolated, and whether there are subnetworks within the overall network.

What travels along the links depends on the composition of the network. In the days when international telephone calls required operators, one of the things that traveled along the links between operators was jokes. At least that was what I was told by a friend who used to be an international operator.

I was relating an experience that I had before the days of direct dialing and the Internet. I was in England, due to get on a plane to fly to Australia, and the person sending me off told me a new joke. When I reached Sydney, I was greeted by someone who told me the same joke. I couldn't work out how the joke got there so fast until my telephone operator friend told me that she and the other operators used to fill in slack periods by calling each other and sharing jokes (and other tasty tidbits) that they had overheard in people's conversations.

Whether it's jokes, packages, information, or even a viral disease like influenza that's traveling through a network, mathematicians represent the network in the same way—as a pattern of dots (the nodes) connected by lines (the links). By doing so, they have been able to identify what different networks have in common, and how these features affect the performance of those networks.

Networks, in the eyes of a complexity scientist, are complex systems. (If nodes or links can be added or removed, they can often become complex adaptive systems.) The links between the nodes represent local interactions, but the emergent properties of the network as a whole are somehow greater than the sum of those local interactions. A city, for example, may be thought of as a very complex network of local interactions between its residents. Some of them run boutiques, others clean the streets, work in offices, and buy their lunch from the local deli. Still others deliver supplies to the deli; provide transportation for those who work in the offices; arrange for the availability of water, power, and sewage treatment; manufacture clothing; and educate children. No one specifically directs all of this activity. The city network is a self-organizing system, and usually a successful one.

How does it all work? Mathematicians using line and dot models have come up with some basic rules we can use to understand how the networks we belong to function, and they are beginning to work out how we can use that understanding to help solve everyday problems more effectively.

Connectivity

There are two extreme forms of network. One is laid out in a prescribed pattern, like the wiring diagram of a computer or the hierarchical structure of an army. Such networks are totally ordered, like the regularly positioned atoms in a crystal lattice or the precise patterning of a spider's web.

The other extreme form of network is one in which the links are formed at random, like the crisscrossing streaks of paint in a piece of art created by Jackson Pollock.

Random networks are a mathematician's joy. One of the most famous findings in the mathematics of networks is that if you keep adding random links to a sparsely connected network, there comes a point when *suddenly* the whole network is interconnected. A Pollock painting might have some disconnected areas, for example, but one more streak of paint could mean that you could suddenly get from one crossing point to any other along the streaks that link the crossing points. Likewise, if you have a group of people who mostly don't know each other, it is impossible to get a rumor spreading within the group. If all those in the group know and talk to just one other person, though, a rumor can spread like wildfire. This is because the point at which interconnectedness suddenly occurs is when there is an average of *exactly* one link per node.

Random networks may be a mathematician's joy, but there are comparatively few real-world examples. One is the highway system; cities are the nodes, and highways are the links, and the mathematical distribution of links follows the classic bell curve that is characteristic of a random distribution. One might also think that a random network of disease is initiated when someone sneezes in public and passes the disease on to a group of strangers, who in turn pass it on to other strangers through their own sneezing. In practice, though, the resultant network of infection is far from random, as I'll show later.

Most networks in real life are somewhere in between completely ordered and totally random. It was the idea of a totally random network, though, that sparked psychologist Stanley Milgram's famous small world experiment, which involved trying to send a letter to a perfect stranger.

Milgram wanted to know how many links there might be in a chain of connectedness between two random strangers. To test his idea, he performed a series of experiments that have been copied and adapted many times since, in which a document is sent by a stranger to a target person via a chain of acquaintances who are on first-name terms with those on either side of them in the chain.

In his best-known experiment, Milgram asked a random group of 196 people in Nebraska and 100 people in Boston to try to get a letter to a stockbroker in Boston by sending the letter to someone they knew by their first name and who might be closer to the target, with a request to send it on to someone that *they* knew by their first name and who might be closer still, and so on.

Sixty-four chains reached their target, with an average number of just 5.5 steps for those originating in Nebraska and 4.4 for those originating in Boston. Milgram's experiment provided the inspiration for John Guare's 1990 play *Six Degrees of Separation*, which explores the idea that we are "bound to everyone on this planet by a trail of six people" and popularized the phrase, providing the stimulus for the many plays, books, films, and TV shows that have since been based on the same theme.

The simple statistics of a random network provide a plausible reason for why short chains of connection might be the norm. Let's suppose that each of us knows 100 people fairly well, and that each of *them* knows 100 people fairly well. So in just two links, any of them will be connected to every other one of them. That's 10,000 people within just two links of each other. If each of *them* knows 100 people, that's 100 x 10,000 = 1 million people within three links of each other. Keep carrying the argument forward, and by the time you get to six links that's a thousand billion people, which is considerably larger than our estimated world population of around 7 billion.

Other networks have similarly short chains, although the numbers are slightly different. The Web, for example, has nineteen degrees of separation, which means that any website is an average of nineteen clicks from any other. This may seem like a lot, but it's a small number compared to the billion plus pages now on offer. If the links between pages were random, the figure could be accounted for by an average of just three links per website, since one billion is approximately equal to 3^{19}.

The six degrees experiment and the analysis of the Web demonstrate the reality of the small world hypothesis, although some scholars have discovered serious flaws in Milgram's work that make his conclusions more tenuous. In neither case, though, do the results *prove* that our global social network or the network of websites are interconnected in a purely random way. There could be many other ways of connection that would lead to similarly short chains.

When Duncan Watts and Steve Strogatz from Cornell University took a closer look at the problem of connectedness, for example, they discovered that our social worlds could consist of tight, intimate networks yet still display the small world phenomenon of six degrees of separation.

The trick was to use a different sort of connectivity. Instead of many random links in an otherwise unstructured network, just a few random links between members of different social networks created the same effect.

My father took advantage of this effect when he was searching for a special tool for his jeweler's lathe. No one in his local group had one, but one of them suggested contacting a friend in England who might. That friend didn't have one either, but through *his* local network he found someone who did, and the tool was duly shipped to Australia. Problem solved, thanks to one long-range link between two otherwise isolated social groups.

What's surprising is how few such links are needed to shrink a world of otherwise largely isolated social networks. They provide shortcuts between nodes that would otherwise be many links apart. In mathematical terms, in the realm between completely ordered networks and completely random networks, there is a large region

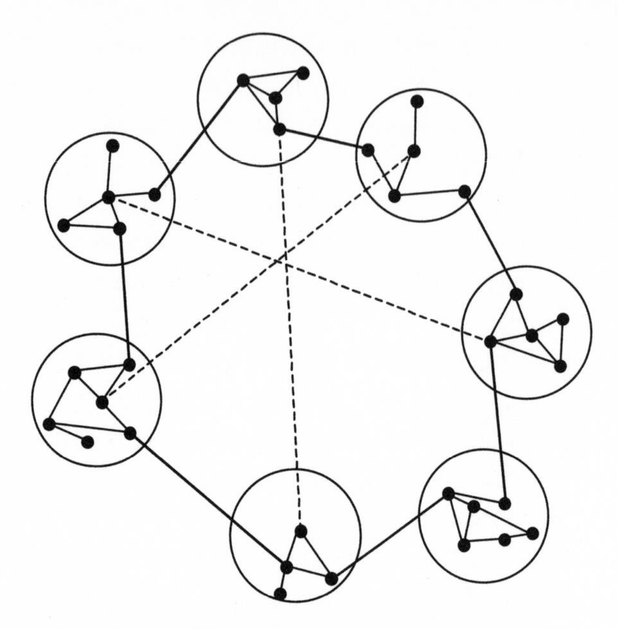

Schematic representation of a set of seven small world networks (black dots linked by full lines, with circles emphasizing each network). Some nodes are separated by as many as twelve links. With the addition of just three long-range links (dotted lines), the number of links separating any two nodes is dramatically reduced.

where groups of nodes form tight clusters, but where there is only a small number of steps between nodes in different clusters, thanks to the presence of a few random links between the clusters.

This is a region that is characteristic of the real world. Friendship networks, for example, can exhibit tight clustering—friends of any particular person are also likely to be friends of each other, yet the average number of friendships in a chain connecting two people in different networks can still be very small.

My wife discovered just *how* small the number could be when she went to lunch with some friends from our neighborhood. The conversation turned to mutual acquaintances, and Wendy mentioned how surprised she had been when we moved to the area and learned that one of our neighbors had been the headmistress at her English boarding school. She wasn't half as surprised as her host, though, who revealed that his first wife had been to the same school. Then it was Wendy's turn to be surprised again when the wife's name was mentioned—she had not only been to the same school but had also been in Wendy's class! The rest of the conversation was taken up with close inquiries by the present wife as to just what the first one had been like.

This sort of connectivity is not so exceptional, and many of us can doubtlessly recall similar stories. When Watts and Strogatz tested their ideas on different networks—which included those of Hollywood actor Kevin Bacon, the neural network of the nematode worm *Caenorhabditis elegans*, and the electrical power grid of the western United States—they found in all cases that there were clusters of strong connectivity but that the network as a whole still constituted a small world in which any node was just a few steps away from any other.

That's not to say that it is easy to discover the shortest path.

Just as this book was going to press, Duncan Watts advised me of recent work in which he and his colleagues have distinguished between two versions of the small world hypothesis. What they call the "topological" version holds that for a randomly chosen pair of individuals in a population there is a high probability that there will be a short chain of connection between them. The stronger "algorithmic" version claims that ordinary individuals can navigate these chains themselves. There is strong evidence for the former, but the evidence for the latter is much more tenuous.

It is especially important to distinguish between the two because each is relevant to different social processes. "The spread of a sexually-transmitted disease along networks of sexual relations," they say, "does not require that the participants have any awareness of the disease, or any intention to spread it [and they] need only be connected in the topological sense. . . . Individuals attempting to 'network' [however] must actively traverse chains of referrals [and] thus must be connected in the algorithmic sense."

In other words, networking can involve a lot more than six degrees of separation because we may enter many blind alleys in the process and have to backtrack or try again, but the spread of diseases such as swine flu does indeed happen via six degrees of separation.

Most of the letters in Milgram's experiment never reached their target because someone, somewhere along a chain, simply couldn't be bothered to send it on. When Watts and his coworkers adapted Milgram's experiment to the world of the Internet, they found something similar. Of the 24,163 chains that were started, only 384 reached their target—just 1.5 percent.

When participants didn't send the message on, it wasn't because they couldn't think of anyone to send it to. When asked, most of them gave lack of interest or lack of incentive as the reason. So that's a lesson for anyone wanting to establish a chain of contacts to someone they don't know. There will be little chance of success unless the person starting the chain is able to provide an incentive that can be passed along the chain.

Of course, chain letters (nowadays largely superseded by chain e-mails) often provide the incentive of monetary rewards or the threat of bad luck to persuade people to keep the chain going. One notorious example of a chain letter from the 1930s was the "Prosperity Club" (also known as "Send-a-dime"). Here is the wording of the actual letter, which tells it all:

> *Prosperity Club.*
>
> *We are sending you a membership to the prosperity club so be sure and send five letters, as this chain has never been broken.*

1

Prosperity Club.

We are sending you a membership to the prosperity Club so be sure and send five letters, as this Chain has never been broken.

In God we trust.

Mrs D. O. Ostby Sheyenne N.D.
Miss Mary Borthwick
Warwick N.D.
Miss Alice E. Kennedy
Sheyenne, N.D.
Miss Bertha A. Jacobson
Sheyenne N.D.
Miss Magnhild Larson
Sheyenne N.D.

2

Oscar Hosum Madock N.D.
This Chain was started in hope of bringing prosperity within 3 days, make 5 copies of this letter leaving off the top name and adding your name at the bottom and mail to 5 of your friends to whom you wish prosperity to come in Omitting the top name send that person 10¢ (wrapped in paper) as a Charity donation.

3

In turn as your name leaves the top you will receive 15,625 letters with donations amounting to $1562,50.

Is this worth a a dime of your money? have the faith your friends have had and this Chain will never be broken,

In God we Trust.
Mrs. D.O. Ostby, Sheyenne N.D.
Miss Mary Borthwick, Warwick N.D.
Miss Alice E. Kennedy, Sheyenne, N.D.
Miss Bertha A Jacobson, Sheyenne N.D.
Miss Magnhild Larson, Sheyenne N.D.
Oscar Hasum, Madock N.D.

This chain was started in hope of bringing prosperity within 3 days. Make 5 copies of this letter leaving off the top name and adding your name at the bottom and mail to 5 of your friends to whom you wish prosperity to come in. Omitting the top name send that person 10 ct (wrapped in paper) as a charity donation.

In turn as your name leaves the top you will receive 15,625 letters with donations amounting to $1562.50.

Is this worth a dime of your money? Have the faith your friends have had and this chain will never be broken.

The effectiveness of this particular letter, which started in Denver, Colorado, in 1935, was such that the Denver Post Office was swamped with hundreds of thousands of letters, which spilled over to St. Louis, Missouri, and other cities. Modern equivalents include chain e-mails on platforms like Facebook, YouTube, and MySpace, and even chain text messages on cell phones.

Chain letters and their successors offer substantial rewards to keep a chain going or severe punishments for breaking it. Watts and his coworkers found that it doesn't take much incentive to keep a chain going but that it does usually take some. Just the feeling that an e-mail has a chance of reaching its target is sometimes sufficient, as shown by the fact that 44 percent of all completed chains were to a professor at a well-known university who was an "obvious" target to the largely middle-class participants in the United States, compared to the other seventeen mainly overseas targets.

Even when there is an incentive to extend a chain, we still need to find some efficient way of making it manageably short. One mathematically based method relies on forwarding the message to the contact who is closest to the target in terms of lattice distance (separated from the target by the smallest number of links). This is all very well if one has an Olympian perspective of the whole network, but it is hardly practicable in real social terms.

A more realistic method, which combines knowledge of network ties and social identity, relies on the fact that we are all members of numerous networks. We can take advantage of this fact by sending our message to the contact who seems closest to the target in terms of social distance—in other words, whose social position means that they have the highest chance of knowing the target, or knowing someone who does.

This is what the participants in Milgram's experiment and its successors seem to have done intuitively. It was also what happened when my father was searching for a lathe tool. He found it very quickly because his friend identified the right person to contact—one who was the closest in social distance to the problem.

When someone in my family wanted to make contact with a distant relative, though, they used a different approach—they asked Auntie Lilla.

Auntie Lilla had links with just about everyone in our family network. When she didn't have a direct link, she certainly knew someone who did. By asking Auntie Lilla, you could save yourself a lot of hassle in searching for relatives. Usually it just took one telephone call.

In network terms, Auntie Lilla was a hub—a node with links to many other nodes. The importance of hubs in large-scale networks has only been realized since the early 2000s.

Hubs

Hubs emerge as a consequence of network self-organization. When mathematician Albert-László Barabási and his bright student Réka Albert looked at the distribution of links in the actor networks and the power grid that Watts and Strogatz studied, they expected to find a classical bell curve. Such curves have the shape of a hanging church bell, the peak corresponding to some average number of links. Instead, they found that a few nodes had many connections (far more than could be reconciled with the Watts and Strogatz model) while the rest had many fewer connections.

There are a number of ways to express a power law. It is akin to Murphy's Law of Management, which states that 80 percent of the work is done by 20 percent of the employees, or the customer service law, which says that 80 percent of the customer service problems are created by 20 percent of the customers.

These can be visualized by drawing them as a graph (see upper figure page 119). When this is done for the distribution of global wealth, the graph takes the shape of half of the bowl of a champagne or daiquiri glass, and the 80:20 rule becomes what is known as the "champagne glass" effect. The graph reveals more detail than the stated rule. Produced as part of a U.N. Human Development Report, it shows that the richest 20 percent of humanity hoards 83 percent of the wealth, while the poorest 60 percent subsists on 6 percent of the wealth. The champagne glass shape of the graph denotes a power law, rather like Newton's Law of Gravity, which states that the force of gravity F between two objects is inversely proportional to the square of the distance d between them. This is written as $F{:}d^{-2}$.

When Barabási and Albert looked at the distribution of links in Kevin Bacon's actor network (see lower figure page 119), they found a very similar law. Using $P(k)$ as the probability that a node would have a certain number of connections k, they found that $P(k) = k^{-2.3}$.

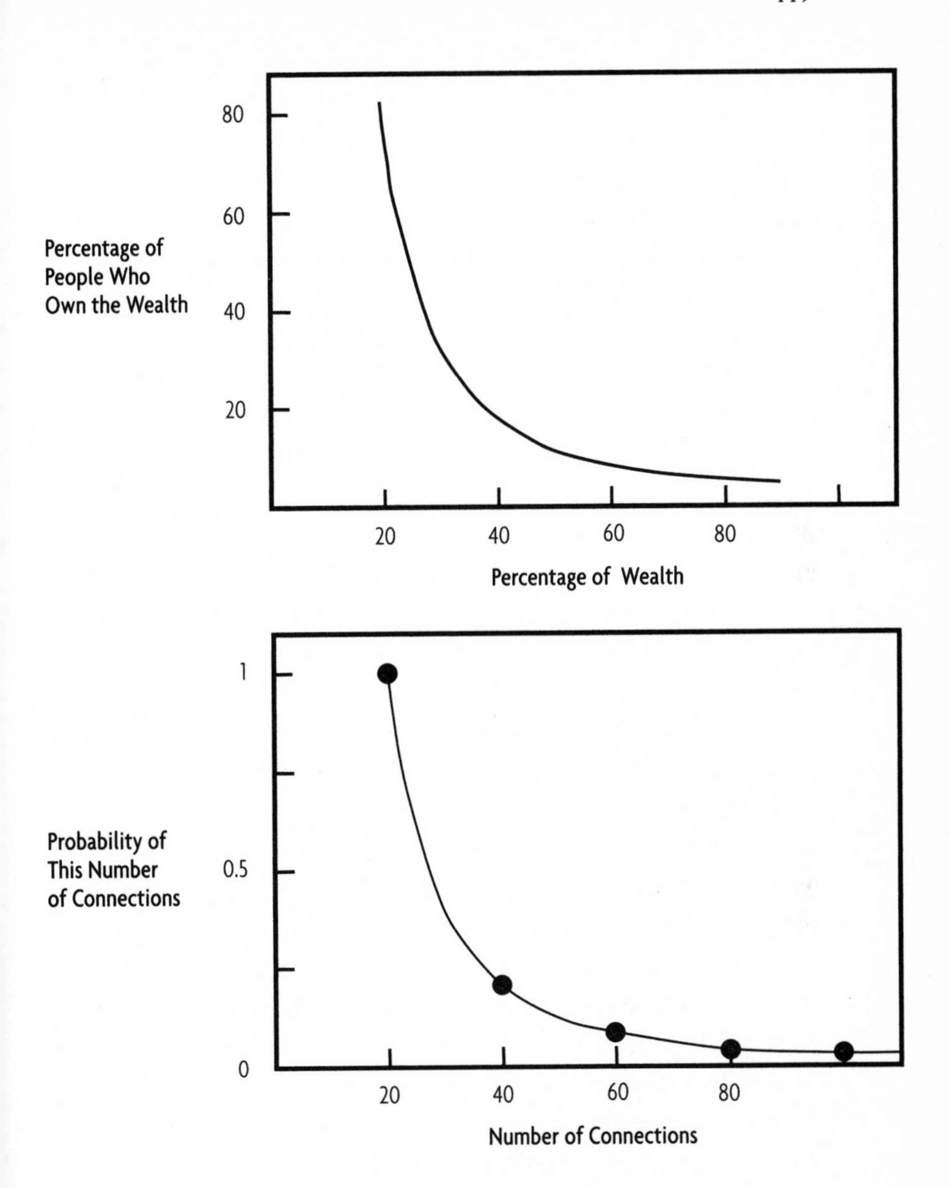

(upper) Power law distribution of the world's wealth.
(lower) Power law distribution of connections in Kevin Bacon's actor network. The exponents in the two graphs are very similar, −2.3 for Kevin Bacon, and −2.1 for the distribution of wealth.

Actors are not known for their gravity, so they sought for some other explanation as to why the distribution of links should follow such a power law, and why similar power laws are found in so many other networks, such as the Web. Other examples that have since been discovered include food webs* in nature, transportation networks, corporate coownership networks, collaboration networks of scientists, and boards of directors. Interestingly, power laws have also been discovered in the network of protein interactions in the cell, genetic regulatory networks, and neural networks in the brain.

Most of these networks are open in the sense that new nodes and links keep on being added. They are, in other words, complex adaptive systems. New actors keep entering the Hollywood network. New routes get added to the airline network. New pages are continuously being added to the Web. New research papers are constantly being added to the scientific literature. Such networks are less like a carefully constructed fisherman's net and more like Topsy in *Uncle Tom's Cabin.* When asked if she knew who made her, Topsy replied, "I s'pect I growed. Don't think nobody never made me."

When new nodes are added to a network, they make links with those already present. Barabási and Albert wondered whether these links might follow the Matthew effect, which takes its name from the verse in the biblical book of Matthew: "For unto every one that hath shall be given, and he shall have abundance" (25:29).

The Matthew effect highlights the fact that new nodes are most likely to link with nodes that already have a large number of links. Thus, as Barabási and Albert explain: "A new actor . . . is more likely to be cast in a supporting role with more established and better-known actors . . . , a newly created Web page will be more likely to include links to well-known popular documents with already high

---

* *Food webs* describes this system more accurately than *food chains.*

connectivity, and a new manuscript is more likely to cite a well-known and thus much-cited paper than its less-cited and consequently less-known peer."

Barabási and Albert's analysis revealed that the Matthew effect would produce a power law distribution of node connections. We still don't know whether real-life social networks grow in this style, as there are many realistic ways in which the power law distribution of connections in a growing network might be accounted for.

Barabási and Albert knew this, and they recognized that their seminal paper was just a first step, using a very simplified model. In reality, networks develop over time not just by adding new nodes and links but by adding new links between existing nodes, rearranging the links, and sometimes even losing them. In our own social networks, for example, we make new friends who are friends of others in our network. Friendships may also wax and wane, or even be lost altogether. Friends may also move away or, sadly, die, ceasing to be nodes in the network.

All of these possibilities have been incorporated into the burgeoning crop of network models that has been stimulated by Barabási and Albert's work. One particularly interesting idea is the Darwinian notion that nodes have some sort of evolutionary fitness, so that the fitter nodes attract more links. In these terms, my Auntie Lilla was extraordinarily fit, partly because of her outgoing personality, and also because people knew that she had many links to others.

This combination neatly fits into Barabási and Albert's original scheme by saying that preferential attachment is driven by the product of a node's fitness and connectivity. The idea of fitness has had many further ramifications, including the fact that under some circumstances just one node can grab *all* of the links. The network then takes the form of a star with one central hub and no rim. Barabási advanced Microsoft's Windows operating system as an example of such a hub, which it still is. Another example that springs to mind is

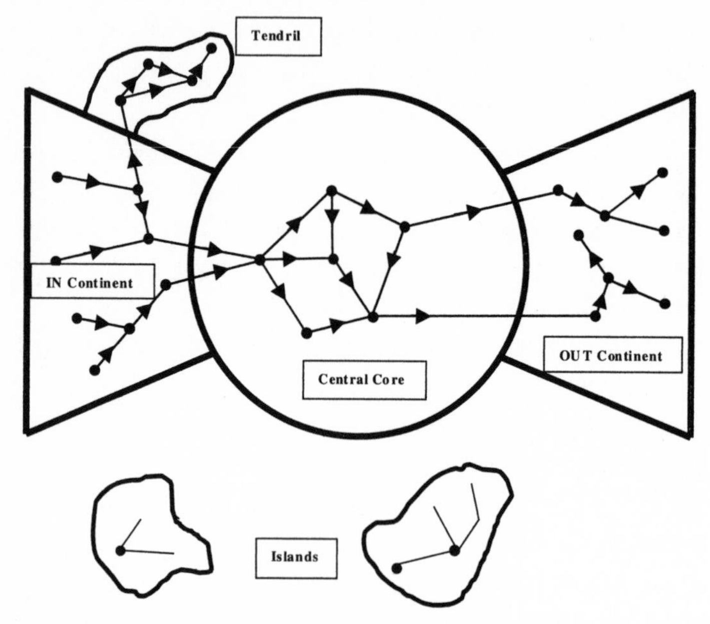

Simplified bow-tie network model, with arrows indicating unidirectional links between nodes (black dots).

that of military censorship, which requires that all letters have to be sent to the censor for examination before being passed on, with previous direct personal links between friends being extinguished by the introduction of the censor into the network.

Another major discovery about networks is the role of directedness, the degree to which the connection along a link goes one way or both ways. A food web in nature is directed; in a link between a bird and a worm, the bird eats the worm, never the other way around. Links are often one-way on the Web. You might link your personal web page to a popular site, but very seldom will the site link back to you.

If the traffic along links is one-way rather than two-way, then the whole idea of six degrees of separation, or *any number* of degrees of separation, goes out the window. The network breaks up into four continents—a central core, an (amusingly named) in-continent, an out-continent, and a group of isolated islands.

The shape of this image has led to the name "bow tie" theory. One look at it shows that we can *never* reach any part of the in-continent from any other region of the network. Indeed, there are many nodes in the in-continent that cannot be reached from other nodes in the same region.

Once we are in the out-continent we can never go back, and once we are on an island we can never escape.

These strictures apply to any network made up of one-way links. Only because the Web has some two-way links are the strictures relaxed, and navigation becomes possible over a wide area. Even so, there are many areas that remain inaccessible unless one can specify the exact web address.

A significant property of power law networks is that they are very stable to a random loss of nodes; if a few are lost or damaged, the performance of the network is not significantly affected.

Only when a major hub is knocked out does real damage occur. One consequence for natural ecosystems, for example, is that the loss of one keystone species can trigger the collapse of the whole system. Unfortunately, we do not always know or recognize which particular species fills the keystone role.

The failure can take the form of a catastrophic cascade, such as large-scale blackouts when the loss of a major power station produces overloads on adjacent power stations and lines. These collapse in their turn, causing even bigger problems further down the line. The ongoing cascade can affect a very large area indeed, as it did in the Northeast Blackout of August 14, 2003. This blackout encompassed major parts of the northeastern and midwestern United States as well as Ontario, Canada, and was at the time the largest blackout in history.

Pass It Along

Our current picture of real-life networks has two main features:

- Many networks have high local clustering and short global path lengths.
- Many networks also have a highly skewed distribution of connectivity with respect to different nodes, with a few nodes (hubs) having a large number of connections but most having many fewer connections. Often (but not always) the distribution follows a power law.

This understanding has led to important advances in the way we think about two things in particular: the spread of disease and the spread of information. They can be thought of as the same thing if we see ourselves as becoming infected with new information and infecting someone else when we pass the information along.

Networks of infection used to be thought of as largely random, like the picture I presented earlier of someone sneezing in public and passing the infection on to a group of random strangers. The new picture of networks, though, has given rise to radical rethinking about how viral diseases like influenza and HIV/AIDS spread so effectively. They do it by using two elements of a network: shortcuts and hubs.

A signal example of the way in which diseases spread is provided by the story of how the Black Death made it to the English village of Eyam in 1665. The network shortcut in this case was a link between the village tailor, George Viccars, and his London supplier, who sent Viccars a bundle of flea-infested cloth. The fleas were vectors for the Black Death, and Viccars was dead within a week.

In the meantime, though, he had acted as a hub to spread the disease to many others in the village. The village itself could have acted as a hub to spread the disease to surrounding villages were it not for the leadership of the rector and the minister, who persuaded the villagers to quarantine the entire village, allowing no one in or out. Around 540 of the village population of 800 died, but the surrounding villages were unaffected.

In modern times, the Black Death has been replaced by HIV/AIDS and other sexually transmitted diseases and pandemic influenza. The HIV/AIDS hubs are people who are particularly sexually active; some have even been known to deliberately affect partners. Pandemic influenza is another killer disease, and in this case the hubs are places where people congregate, such as schools, concert venues, airports, and cruise ships.

Mathematical analyses show that it is much more effective to concentrate on the hubs than on the individuals in the network when it comes to controlling the spread of disease. In the case of influenza, that mathematical advice is being taken by many governments. During the 2009 outbreak of swine flu, for example, if one child in a U.K. school was diagnosed with the disease, the whole school was closed down immediately until the danger of infection was past. In Mexico, the origin of the disease, large public gatherings were banned.

Removing the hubs of sexually transmitted diseases is a much more difficult affair, impinging as it does on the balance of human rights between the infector and his or her partners/victims, and also on the difficulty of identifying the hubs. Education is certainly part of the answer, although it can sometimes go awry, as a colleague of my wife found out when she gave a talk on venereal disease to some children in an African school. She showed a film to demonstrate the chain of events that take place when one person gives the disease to another. At the end of the film one of the children asked, "If I give it to someone else, does that mean that I won't then have it myself?"

It turned out that every child in the room thought the same thing: that you got rid of a sexually transmitted disease by passing it on. A friend who is an aid worker in Africa tells me that this is also a common misconception in the adult population, and that people who suffer from AIDS or other sexually transmitted diseases often believe that they can get rid of the diseases by passing them on.

If a person makes repeated efforts to get rid of the disease in this way, they become a hub for its transmission. But hubs aren't the only problem when it comes to the spread of disease. As Watts and Strogatz's work has revealed, a large world can shrink to a small one with just the addition of a few long-range random links. Identifying those links is an almost insuperable problem, which is part of the reason why large-scale vaccination continues to be used, even though it is like using a sledgehammer to crack a nut.

The best thing we can do as individuals to prevent the spread of epidemic diseases is to keep away from hubs. It would also be very helpful to keep an eye open for anyone (including ourselves) who might be acting as a long-range link. Many people who visited Mexico did this unconsciously during the swine flu epidemic (probably out of fear rather than community consciousness) when they had themselves checked out by a doctor as soon as news of the epidemic appeared.

The spread of information poses the opposite problem: how to keep the hubs from failing. We want to maintain the hubs because these keep the network stable and intact. When my Auntie Lilla died, for example, it wasn't just that we couldn't ask her questions any more. Without her, the family communications network seemed to disintegrate.

We also want to maintain or establish long-range links, because these make our networks manageable through the small world effect. I used to live in an English village, and it had a highly interconnected social network. If someone relayed a juicy bit of gossip, it would be around the village in a matter of hours.

It could take days for the gossip to make its way to a nearby village, though, because villages tended to keep to themselves and there were very few links between them. Then someone moved to the next village who had a close friend in our village. This provided the necessary long-range link, and the gossip suddenly spread to the next village just as rapidly as it did to our neighbors.

The link arose spontaneously, as part of the ongoing self-organization of the network. Hubs also arise spontaneously in any network that obeys a power law, whatever the underlying mechanism that leads to the power law might be. How can we use this knowledge of links and hubs to better use our networks in real life?

Influence and Affluence

The above considerations suggest that there are three main ways to use networks more effectively.

1. Speed up the transmission of information or anything else that is being passed along a network by identifying and using long-range links or establishing new ones.
2. Provide an incentive for people to keep a chain going once it has started.
3. Identify and make use of the hubs that exist in most real-life networks.

The use of hubs looks like the best bet at first sight, because these have the largest number of links to other nodes in the network. Duncan Watts and his colleagues have discovered, however, that hubs may not be quite as important as they appear to be at first sight.

Nodes may have made links to the hub because they see the hubs as important—as centers of power or centers of influence, for example.

This is why people send petitions to politicians. It is why I sent a copy of my last book, about game theory, to Al Gore—he is a hub in the ongoing debate about climate change. An understanding of the problems revealed by game theory could really help him, but of course I didn't get to first base—not because Gore doesn't care but because (as his secretary politely pointed out) he receives endless requests for help and endless offers of help. If he sees my book, it will only be by accident.

My setback points up the major problem in trying to use people of influence as hubs to press your case. People of influence they may be, but their status as such also means that they are less easily influenced than most, partly because of time constraints but mainly because such people have their own firmly worked out agendas and their own definite methods of dealing with those agendas.

A theory from the 1950s, promoted by Elihu Katz and Paul Lazarsfeld in their book *Personal Influence*, suggests that the role of opinion leaders in the formation of public opinion is that information (so-called) provided by the media "flows" to opinion leaders, and thence to their band of followers. They call the opinion leaders "influentials" and argue that influentials act as intermediaries between the originators of ideas and the majority of society.

Marketers in particular have adopted this theory and use it in planning their campaigns. Recent work in network science, however, has largely debunked it. It now seems much more likely that if we want to kick off a cascade of influence through a community, our best bet is to forget about influentials and concentrate on persuading a critical mass of individuals.

Computer simulations have revealed that such cascades can be of two types. First, they may be local, as when my network of friends and their friends kindly get excited about a new book of mine and pass the word around, producing a flurry of sales. Second, they may be global, as happens with large-scale bestsellers such as the Twilight or Harry Potter series that suddenly seem to catch the public imagination.

The difference, according to simulations, is that global cascades occur only if there is a critical mass of individuals who become enthusiastic about a book after a single mention of it by someone else who has become enthusiastic about it. The global cascades that produce crazes, such as the hula-hoop craze of the 1950s, have a similar origin. Once enough people had seen someone else with one, and wanted one for themselves, the craze took off.

The idea that a critical mass of "early adopters" (people who take up something after a single exposure) is needed to start a cascade of acceptance isn't just confined to crazes. It equally underlies the diffusion of innovations, and their ability to "cross the chasm" from innovation to success. It's not so easy to get a critical mass going, though, as the history of viral marketing reveals.

Traditional marketing is targeted at individuals. Viral marketing aims at getting people to spread the message through their networks, in the manner of a virus infecting a computer network. Of course, computer viruses are a lot easier to spread than advertising.

Traditional marketers have two basic options for increasing their sales. They can make their product more attractive, so that there is a greater probability that individuals will buy it, or they can use advertising campaigns to increase the number of individuals who receive their message.

Viral marketers have a third option: getting their message to reproduce by word of mouth or via the Internet. If the reproduction rate is greater than unity (each individual passes the message on to more than one new person), then the message (and the product) will grow and flourish. Otherwise it will die away.

The catch is the same thing that makes sending a letter along a chain so difficult: individuals must be motivated to pass the message on. This means that they must be sufficiently happy with the product to *want* to pass the message on.

Setting up the circumstances for this to happen is no easy matter. Network science pioneer Duncan Watts and his collaborators reported a "contagious media" contest conducted by the media-art nonprofit organization Eyebeam.org, in which a roomful of subject matter experts failed to predict which of sixty submitted websites on their subject would generate the most page views. They concluded from this and other evidence that it is "extremely difficult, and perhaps impossible to consistently create media that will spread virally from a small seed to millions of people."

Book publishers would surely agree. If it were possible to predict a bestseller, that is all they would need to publish. The history of many bestsellers shows how difficult it is to make such a prediction. *Zen and the Art of Motorcycle Maintenance*, for example, was submitted to 121 publishers before William Morrow accepted it. The publisher accepted it not because it thought the book would make money but because the book ought to see the light of day. It was rewarded with worldwide sales in excess of 4 million copies.

Zen and the Art of Motorcycle Maintenance virally marketed itself, but Watts' study shows how extraordinarily difficult it is to make this happen deliberately. With a different group of collaborators, he investigated why this might be in the competitive music download market.

Watts and his colleagues got more than fourteen thousand teenagers to take part in an experiment in which they rated unknown songs by unknown bands, either with or without knowledge of the other participants' choices. They were then given the opportunity to download the songs for their own collection. Before reading on, see if you can figure out what the effect might have been on the download rate for a particular song, knowing how other participants had rated it.

Wrong! If you predicted any correlation at all, you were wrong. The only effect of knowing how other participants had rated a song was to increase both the inequality and unpredictability of success. The only correlations were that the highest-rated songs rarely did poorly in the download stakes and that the lowest rated songs rarely did well, but any other result was possible.

As a result of both practical studies and computer modeling, Watts and his colleagues suggested that viral marketers should be content with a much more modest ambition—to establish a "big seed" at the beginning, so that many extra people will still get the message even if the reproduction rate is less than unity as the message is passed on. Instead of trying to start a forest fire from a single spark, set a light to a modest-sized area of forest so that at least the ring of forest around it burns before the fire dies away.

I decided to try this approach with this book, using free articles, practical hints, and other tidbits as the incentives to establish a big seed. The experiment is just starting at the time of writing. It will be interesting to see how it has panned out by the time the book is published.

Synopsis

To sum up the ideas raised in this chapter, there are many ways to use networks, including:

- Setting up a big seed or a critical mass
- Providing an incentive to keep a chain going
- Making your personal node fitter for its purpose
- Helping to establish more two-way links within the network

All of these involve setting up a network not by decree from above but by providing the circumstances under which the network can form, grow, and adapt itself to changing conditions spontaneously, using local interactions as a driving force. My father did it by intuition in establishing and using his own small network. With the recent advances in understanding that I have summarized in this chapter, it is now possible to deliberately plan effective uses of the much larger networks in which most of us are now involved.

EIGHT

Decision Rules OK?

In his famous Australian short story "The Golden Shanty," Edward Dyson tells the tale of an outback pub near a newly discovered goldfield where the owner is unknowingly sitting on a fortune. The bricks of the pub are made of the local gold-bearing clay, a fact that itinerant miners have realized and that leads them to start stealing the pub walls, brick by brick. Only when the landlord accidentally breaks a brick and sees the shining specks of gold within does he realize the true situation.

Those miners discovered gold in the bricks by keeping their eyes open for the obvious. Mostly, though, miners were forced to collect rocks, nuggets, and gravel indiscriminately from a creek bed, and then swill them around in a pan until the gold separated out.

The most experienced prospectors also had a third string to their bow; one that is still used by mining companies the world over. It was to take a distant view of the field and use their experience to look for patterns that might indicate a rich vein.

The old-time prospectors were pretty successful and cleaned out goldfields like the one at Sofala, near where I live in Australia, within a few years. These days, the only gold you are likely to find in the vicinity of Sofala is the Sofala Gold ice cream that's sold in nearby Bathurst. It is a wonderful concoction of vanilla ice cream with lumps of crunchy honeycomb throughout. When I introduced my grandson to it, he used the techniques of the old-time prospectors to extract his favored honeycomb. First, he picked the obvious bits off the surface, then he poked around in the ice cream with a spoon to find more. Finally, he realized that there was a pattern—most of the lumps had settled to the bottom, and all he had to do was to dig underneath the ice cream to find them in high concentration.

We can use the same three techniques when it comes to digging information out of the complex mountain of data that threatens to engulf us every day. First, we can look for the nuggets and precious stones that lie on the surface—the facts that are easy to recognize, the simple decision rules that bypass the complexities of a problem.

Second, when we need to consider complexities, we can collect all of the relevant material and swill it around in the gold pan of our minds until patterns begin to form and nuggets start to separate out. Our aid in this endeavor is classical decision theory, which provides rules for making decisions based on the patterns that emerge and the nuggets of information we glean.

Finally, we can stand back and look for patterns in the unsorted mass of data. Some of these can be uncovered using the modern science of data mining, which reveals trends and patterns that can be veins of pure gold when it comes to making decisions. But beware—other patterns will consist of fool's gold, meaningful at first sight, but on closer examination, worthless. We can sometimes tell the difference between the meaningful and the meaningless with the aid of Ramsey's theorem, which I discuss in chapter 9. Often, though, we still have to rely on our intuition and judgment.

Homo omnisciens—Not

One of my favorite cartoons features an angry husband saying to his wife, "I assume, then, that you regard yourself as omniscient. If I am wrong, correct me."

If we were really omniscient, we would always reach the best conclusions from the available data. We would take in all the facts, weigh all the options, consider the pros and cons, and come to an opinion on the best course of action.

This is what Benjamin Franklin attempted to do with what he called "moral algebra," which he described in a letter to the English chemist Joseph Priestley.

> My way is to divide half a sheet of paper by a line into two columns; writing over the one Pro, and over the other Con. Then, during three or four days consideration, I put down under the different heads short hints of the different motives, that at different times occur to me, for or against the measure. When I have thus got them all together in one view, I endeavor to estimate their respective weights; and where I find two, one on each side, that seem equal, I strike them both out. If I find a reason pro equal to some two reasons con, I strike out the three. If I judge some two reasons con, equal to three reasons pro, I strike out the five; and thus proceeding I find at length where the balance lies.

The unstated assumption behind Franklin's procedure is that the more information we have, the better our decisions are going to be. He would have been mightily surprised to find that *less* information can often lead to more accurate decisions.

Consider one of the many situations examined by German psychologist Gerd Gigerenzer in his studies of the less-is-more hypothesis. Gigerenzer and his colleague Daniel Goldstein asked students at the University of Chicago whether San Diego or San Antonio had more inhabitants. Only 62 percent plumped for the correct answer of San Diego. When the researchers asked students at the University of Munich in Germany the same question, however, *all* of the students got the answer right.

They got it right because they had *less* information than the American students. It transpired that most of the German students had heard of San Diego, but few of them had heard of San Antonio. With nothing else to go on, they assumed that the city they had heard of probably had more inhabitants.

The German students reached their conclusions by following a rule of thumb. The rule in this case was this: *If you are given a pair of alternatives to choose from, and you recognize only one, choose that one.* Such a rule is known technically as a *heuristic*—a simple rule or set of rules for making good decisions from limited information or in a limited time span. According to Gigerenzer, we use heuristics so often that a more apt title for our species would be *Homo heuristicus.* Gigerenzer and his colleagues have shown that even when there is a mountain of information available, *Homo heuristicus* can often out-perform *Homo sapiens* by concentrating on the nuggets, rather than on the whole pile of dirt.

Franklin's approach to decision making was a heuristic, albeit a rather cumbersome one. A modern equivalent to Franklin's approach is force field analysis, which is used by psychologists and business advisors as a tool for helping people understand and achieve change.

Force field analysis is based on the idea that issues are "held in balance by the interaction of two opposing sets of forces—those seeking to promote change (driving forces) and those attempting to maintain the status quo (restraining forces)." People wishing to make a change in their lives are encouraged by a counselor or therapist to prepare a force field diagram with two columns, one containing the driving forces and one containing the restraining forces, drawn as arrows pointing in opposite directions whose lengths represent the magnitude of the force.

A quick look at such a diagram reveals where the major perceived issues lie. A closer look can sometimes show that the real issue has been left off entirely. Someone who wants to lose weight, for example, may list exercise, stress, availability of food, and the like as major factors but completely omit comfort eating, which may be the most important factor of all but is kept out of mind until a therapist asks whether it should be on the list.

In this sense, force field analysis is not so much a tool for action as an aid to insight. When it comes to many of the choices for actions we have to take in life, recent research has shown that simple heuristics can often be just as helpful.

In fact, simple heuristics can sometimes even improve on more complicated approaches. The American students in the example above had much more information at their disposal than did their German counterparts. All of the Americans in the study had probably heard of San Antonio. Most of them would have known that the Alamo is located there, and many would have had other snippets of information at their disposal. They could have put this information into pro and con columns or listed it in force field style, but it is doubtful whether this would have improved their performance much. It surely would not have raised their performance to the level of their German counterparts, who benefited from a sheer *lack* of information.

The cue of recognizability that the Germans used is not the only simple heuristic that is available to us when it comes to making decisions. Gigerenzer and his colleagues have identified a total of ten heuristics, many of which we already use in our everyday lives, and some of which we could make far better use of. I have selected five that seem to me to offer particular promise as simple strategies for helping to resolve the complex problems of life. (The other five are listed in the notes to this chapter on pages 232–234.)

1. RECOGNITION

If you are given a pair of alternatives and recognize only one, choose that one.

The recognition heuristic works best in situations of an intermediate degree of ignorance, as I discovered when I got lost in Bangkok a few years ago.

I don't speak a word of Thai, and I could not remember either the name of my hotel or the street that it was on, so I could have been in considerable trouble. Luckily, my wife and I had spent a fair bit of time walking around the city, and I could remember the names of a few of the principal streets, since these were the ones that we had come across most often. I crisscrossed the area I was in and eventually recognized one of these names. Walking along that street, I soon recognized another name and, with my bearings established, I was able to make my way safely back to the hotel.

If I hadn't remembered any of the names that I had seen, I would have been totally lost. If I had remembered the name of every street that we had traversed, however, I would have been in an equal jam, because I wouldn't have known which one to choose in the absence of other information. By remaining partially ignorant, remembering only the names of the streets we had seen most frequently, I was wholly successful in resolving my dilemma.

Gigerenzer and Goldstein found that the less-is-more recognition principle could even be used to guide investments in stocks and shares. They took the list of companies on the German stock exchange out to the streets of Munich and asked pedestrians which names they recognized. Putting their money where their belief was, they then invested a "non-trivial" amount of their personal savings in a portfolio of the most-recognized companies. After six months, the value of their portfolio had increased by 47 percent—well above the increase of 30 percent for the German Dax30 market indicator over the same period.

There is no particular mystery as to why the less-is-more principle worked in the researchers' favor. The companies most people had heard of were the ones that were making news because they were doing well in the strong bull market of the time. It is unclear whether the same strategy would work in the era of the credit crunch, but a beneficial degree of ignorance of less successful companies could produce comparative bliss in the right circumstances. The crucial point is that we should recognize the significant without being distracted by the confused and messy background of the insignificant.

Another factor is our relative inability to recall old information. If the information is marginal to our usual activities in life, we are more likely to remember new information. A group of Canadian sociologists proved this point in spectacular fashion when they managed to make the name Sebastian Weisdorf famous overnight.

Weisdorf was actually one of their friends, and the researchers included his name in a list of one hundred nonfamous people. After showing individuals the list and telling them that the people on it were not famous, they then showed them another list in which they had mixed forty of those names with twenty new nonfamous names and sixty recognizably famous names (but not so famous that people could usually say what they were famous for).

Most of the people who were shown both lists on the same day could easily identify who on the new list was famous and who was not. Those who were asked on the following day, however, generally identified Weisdorf as famous. His name had become famous overnight because these people had by then forgotten the details of the first list but vaguely remembered having seen the name Sebastian Weisdorf somewhere, and they used this recognition cue to class him among the famous. As a result, Sebastian Weisdorf is now truly famous—among psychologists—for having become famous overnight.

Recognition, then, can be a two-edged sword when it comes to using it as a cue to choose between alternatives. This applies particularly to circumstances in which advertisers have used some of the many tricks they have up their collective sleeve to force recognition on us.

One of those tricks is to include a well-known identity in the advertisement. This backfired recently, however, when American Apparel used an image from the film *Annie Hall* to promote its new line, and was sued by Woody Allen, a star of the film, because "they calculatingly took my name, likeness and image and used them publicly to promote their business."

Recognizability is big business, and it behooves us to act like Allen, and to identify the circumstances under which it may be being used deliberately before we employ it as a primary cue.

That said, I have to admit that using recognition as a major cue may have helped me to save my marriage. I happen to suffer fairly severely from prosopagnosia, which is an inability to recognize faces. I have walked past my own brother on the street without recognizing him, and nearly had a disastrous start to my present marriage when I walked straight past my new wife as she waited to pick me up at a crowded airport.

I was desperately looking around for cues when I noticed her hat. It had a tantalizing air of familiarity, and the few other hats worn by women in the crowd did not strike such a chord. "Aha," I thought, "*she* must be the one that I married."

Luckily, she was. She has never forgotten this incident. For that matter, neither have I. Forgetting, though, can have its benefits.

The advantages of forgetting were first recognized by the pioneering psychologist and philosopher William James. He argued that "in the practical use of our intellect, forgetting is as important a function as recollecting. . . . If we remembered everything, we should on most occasions be as ill off as if we remembered nothing. It would take as long for us to recall a space of time as it took the original time to elapse, and we should never get ahead in our thinking."

James's rationale was nonsense, of course, because it needn't take as long to remember and reexperience something as it did to experience it in the first place. The act of remembering, though, can certainly be enhanced by the ability to forget.

A psychologist friend told me of a waiter in a large restaurant that he frequented who never made a note but always came with the correct dishes and was even able to distribute them to the particular people at the table who had ordered them. Out of professional curiosity, my friend asked the waiter late one evening to use his remarkable memory to remind him of the dishes that he and his friends had eaten. The waiter couldn't remember any of them. When questioned further, he said that as soon as he had served a meal, he wiped the now redundant information from his memory. Otherwise, he said, the information got in the way of memorizing fresh orders.

Forgetting can also help when it comes to using recognition to make choices between alternatives. By forgetting the less desirable options, we are better able to use recognition to pick out the best one when recognition is the primary cue we have to go on.

Even when we recognize all of the alternatives that are available to us, it can still help us in another way. That way is to use fluency—to go with the option that we recognize most easily.

2. FLUENCY

If recognition is all you have to go on in making a choice between alternatives, and you recognize more than one alternative, go with the one that you recognize most easily.

At first glance, fluency looks rather like a recommendation to go with your first instinct, which is the central doctrine of books such as Malcolm Gladwell's *Blink*. In fact, there is a very basic difference between fluency and instinct. Gladwell applied the instinct idea to a wide range of examples, but fluency applies only to situations in which recognition is the main (or only) cue we have available.

If we are looking for a local restaurant on the Internet, for example, we are likely to recognize a few names, but the one that jumps out at us is likely to be the best choice, even if we have never been there. When my wife and I were looking in real estate agents' windows with a view to moving to a new house in our neighborhood, we recognized many of the homes that were for sale, but one in particular caught our eye as being one that we had seen some time ago and rather fancied. When we investigated further, our initial response was confirmed, and we bought it.

The underlying science of the fluency decision strategy is that the alternative that you recognize most easily is also likely to be the most familiar, and therefore (according to the arguments that I presented in the previous section) the most germane to your choice. The strategy appears to go dead against the ancient maxim "Quick decisions are unsafe decisions," but this is because the maxim assumes that we have the option of making a slow decision based on the accumulation of additional information. Fluency is useful in situations where that information is unlikely to be forthcoming, at least during the time available for making the decision. Even when we do have that option, the fluency strategy is often a valuable first guide.

3. TALLYING

Look for cues that might help you make a choice between options, and go with the option that has the greatest excess of positive over negative cues without bothering to try to rate them in order of importance.

Tallying is one of the simplest ways we can use a large number of cues to guide our decisions. It is a shortcut form of Franklin's moral algebra. It consists of drawing up lists of pros and cons for the various options but then going with the longest list and forgetting about such niceties as giving different weights to the different factors.

Charles Darwin used this approach when he was deciding whether to propose to his cousin Emma, whom he had known since childhood. Under the headings "Marry" and "Not Marry" he listed what life might be like in either state as a series of pros and cons. The pros included companionship ("better than a dog, anyhow"), and the cons included the fact that he would have less money for books and wouldn't be able to read in the evenings. The arguments for marriage occupied considerably more space on the page than those against, and Darwin duly married his cousin.

Tallying seems like a rather silly idea, since it takes no account of the relative importance of different factors. Its utility seems so counterintuitive, in fact, that when decision theorist Robyn Dawes demonstrated that it could sometimes out-perform much more complicated approaches, his conclusions were greeted by the scientific community with howls of outrage.

Closer thought, though, shows that tallying is a perfectly reasonable approach when we don't know what weights to give to different pros and cons, or when they all have pretty much equal weight anyway. If we are choosing which of a number of films to go to, for example, we might take into account the qualities of the actors, the names of the directors, where the films are set, and a host of other factors. If we have no idea which of these is more important than another, our best bet is to go to the film with the largest excess of pros over cons.

Even in circumstances that suggest strong scientific reasons for placing different stress on a variety of cues, we are no better off if we don't know what those reasons are. Surprisingly, it need not matter.

When it comes to predicting rainfall, for example, a simple pro and con accounting of factors like the type of cloud, the amount of cloud cover, and whether a cloud was seeded can produce just as accurate a prediction as a more scientific analysis would.

The key here is the word *predicting*. This is what misled Dawes' critics. A scientific, statistical weighting of the different factors is a better fit for *known* data. When it comes to extrapolating those data into the future, though, a simple pro and con list works just as well and (by eliminating the bumps and wiggles in the data) sometimes works even better. When researchers tried it across a range of twenty widely varying categories (including rainfall prediction, mentioned above), Dawes' rule scored an average prediction success rate of 69 percent, while the complex statistical procedure of multiple regression scored 68 percent.

It is worth giving the full list, just to show the range that it covered:

- Attractiveness (of men and women)
- High school dropout rates
- Rates of homelessness in different cities
- Mortality rates in different cities
- City populations
- House prices
- Land rent
- Professors' salaries
- Car accident rates
- Car fuel consumption in different states
- Obesity at age 18
- Body fat
- Fish fertility
- Mammals' sleep
- The amount of oxygen taken up by cow manure
- Biodiversity in the Galapagos Islands

- Rainfall
- Oxidant amounts in Los Angeles
- Ozone concentration in the atmosphere of San Francisco

In each case, the prediction was made from a list of published cues. The attractiveness of famous men, for example, was predicted (without seeing a picture) from their published likeability ratings, the percentage of subjects who recognized the man's name, and whether the man was American.

When we *can* tell that some factors are more important than others, studies have shown that we can do even better by using level-wise tallying, which involves underlining the factors in the pro and con columns that we think are particularly important. If there are more underlined factors in one column than the other, that's our decision made. Otherwise, we cross all those out and underline the ones that we regard as being of intermediate importance. If these still balance out, go back to using the total length of the columns, or toss a coin—it won't matter much, since the decision now depends on minor factors anyway.

Studies have shown that tallying succeeds best when intrinsic predictability is low, the number of cues is within an order of magnitude of the number of choices, and the cues are interdependent. Amazingly, though, we can often do better still with an even simpler method called "take-the-best."

4. TAKE-THE-BEST

When faced with a choice between two options, look for cues and work through them in order of your expectation that they will lead to the best choice. Make your choice on the basis of the first cue that distinguishes between the alternatives.

If you were a female guppy, for example, evolution will have dictated that you prefer males that are the brightest orange. Orangeness would be the first cue on your list in choosing between two possible male partners. If the two males were more-or-less equally orange, though, you would be forced to use a second cue. One that female guppies use in practice is to prefer the male that they have seen mating with someone else.

We use similar cues. Evolution dictates that we are attracted to people whose facial appearance and body shape are symmetrical. We also tend to be attracted to other people's partners. So symmetrical-or-unsymmetrical and partnered-or-partnerless are two cues by which we might choose between potential mates. Other cues might be height, intelligence, income, social status, and so on. Put these in order of priority, start at the top, and as soon as you get to a cue that distinguishes between the two possibilities, stop. That is the essence of take-the-best.

Take-the-best is expected to do well in a wide range of situations in which we have learned through experience or evolution to identify the most reliable cues. Even in the experiment with the twenty widely varying categories, it outperformed all of the other approaches. This is not just a pragmatic result; a careful mathematical analysis, using advanced statistics, has shown that this strategy *should* do well from a purely theoretical point of view. Just by using the most reliable cue first and using the next-best cue only if we have to, we can make the most of our binary choices in life, from choosing between two potential partners to choosing between two possible holiday destinations.

5. SATISFICING

Search through alternatives and choose the first one that exceeds your aspiration level.

Satisficing cuts through the Gordian knot of complexity. Instead of trying to maximize our satisfaction, in the manner of the traditional (mythical) *Homo economicus*, we accept a reasonable level of satisfaction. Instead of cues, we use aspirations to guide our choice between alternatives. We become *Homo pragmaticus*, accepting a reasonable level of satisfaction rather than holding out for the absolute best. What could be simpler?

Well, nothing could be simpler until *Homo greedipus* gets into the act. *Homo greedipus* accepts that it is not always possible to look at every TV on the market before choosing one, or to test every brand of detergent before settling on the best. *Homo greedipus*, though, would like to do better that *Homo pragmaticus* and doesn't like the idea of taking the first reasonable option when there might be something even better just around the corner.

There is indeed a way of doing better. Much better. It doesn't rely on checking out every option before making a decision. In fact, it recognizes that very often this is simply not possible. For example, if you are walking through a crowded flea market looking for a particular item and spot a good deal, chances are that if you keep looking for a better bargain and don't find one, the first one won't be there when you go back. You must make the decision now. When should you buy, and when should you keep looking?

If you are satisficing, your choice is simple. You will have set a reserve price in your mind. If the article is cheaper than the reserve price, you buy. However, your reserve price may be conditional on the prices of the first few examples that you have seen. Maybe, instead, you should turn to the simple and surprising statistics of the secretary problem.

This problem has been cast in various forms (including the search for a bargain), but it is named after the original example, which concerns someone who is interviewing people for a secretary's job. A hundred people have turned up for an interview. The interviewer sees them in random order and must accept or reject a candidate before moving on to the next one. Once a candidate is rejected, there is no going back. What strategy should the interviewer adopt to have the best chance of choosing the perfect candidate?

The answer, discovered by statisticians John Gilbert and Frederick Mosteller in 1966, is that the interviewer should see the first 37 candidates without accepting any of them, and then accept the first candidate after that who is better than any of the first 37. The 37 percent rule gives the interviewer a good chance of finding one of the top candidates, and a 37 percent chance of finding the very best candidate. It can be applied to many situations in life.

In the days before digital cameras, one savvy statistician even applied it while he was on a vacation, when he wanted to take a photo of the most beautiful site to be visited in a remote region but had only one shot left and no opportunity to return to any place once he had left it.

What *Homo greedipus* is doing, of course, is using roughly one-third of the candidates (or photo opportunities) to set an aspiration level and then using satisficing to choose from the remaining two-thirds. Not a bad ploy, one might think, and equally useful in searching for bargains, so long as one can make a reasonable estimate of how many bargains might be available in total. There are other strategies that can do even better, though, as long as we are prepared to relax our standards a bit.

Let's say that we would be happy to find a bargain whose price is in the bottom 10 percent of those at the flea market. We have to guess how many bargains there might be in total. If there may be as many as a hundred, calculations presented by heuristics experts Peter Todd and Geoffrey Miller show that we should look at fourteen without buying and then choose the next one that has a lower price than any we have yet seen. This gives us a whopping 84 percent chance of finding a bargain with a price in the bottom 10 percent—the best we can do.

If we are happy with a price in the bottom 25 percent, Todd and Miller's calculations show that we need to look at only seven before choosing the next one with a lower price. This gives us an even better chance (92 percent) of finding such a bargain.

Those figures are all we really need: 7 percent and 14 percent for bargains in the best 25 percent or best 10 percent. If there are only seven likely bargains in the flea market, these figures mean that we need look at only one before taking the next one with a lower price to have the greatest chance of grabbing the best bargain.

Could we use the same strategy in choosing a life partner? We could, except for two things. One is that we don't know how many potential life partners we are likely to meet. The other is that, while we are looking at someone else as a potential partner, they are also looking at *us*. They may meet our aspirations—but do we meet theirs?

We can probably make a fair guess at how many people we might meet, based on our social patterns and previous experience. We might even be able to manipulate this number by using a dating agency or one of those organizations that helps you to meet people. In any case, the precise numbers aren't all that important. The mathematics shows that even if there were a thousand people in your pool of potential mates, you would still have to meet only thirty of them to set your aspiration level to have a 97 percent chance of finding one in the top 10 percent by subsequently using satisficing.

When you find one of those people, though, they may not be as happy with you as you are with them. Mathematically, this means using your predecision meetings to set your aspiration level at a point where there is a 50:50 chance that the other person (when you find them) will be as interested in you as you are in them. So maybe it's better to confine the math to bargain hunting. It has a better chance of working there.

Decisions at the Edge of Chaos

Sometimes it helps to have our strategies worked out in advance, in the manner of safety briefings on an airplane, so that we can be ready to take appropriate action in rapidly changing circumstances. If we are going to make fast and accurate decisions, we need simple and easily applicable rules to guide us.

In my search for such rules, I came across a seminal study of modern business strategies by Kathleen Eisenhardt from Stanford Univer-

Why Do Simple Rules Work So Well in Resolving Complex Problems?

It is obvious that the rules above work when they are applied to the right sort of problem. But why do they work? The answer lies in the fact that we are using the rules to predict what will happen in the future rather than to understand what has happened in the past.

A simple example will serve to illustrate the difference. Let's say that we want to understand how the atmospheric temperature varies from day to day. If we keep a record of the temperature, a graph of temperature and time will have lots of bumps and wiggles. Some of these will have real significance for our overall picture, but others will be accidental. A sudden dip in temperature, for example, may be the result of a genuine change in the weather pattern or of a chance cloud passing in front of the sun.

We can model the changes in temperature by fitting the observations to a heap of variables like average cloud cover, ground temperature, time of day, and the like. The more variables we put into our model, the better the fit will be, and the more accurately we will reproduce the bumps and wiggles. When it comes to predicting the future, though, what we want is the trends. To predict trends, we need to keep our model fairly simple—that's one reason why the comparatively simple rules mentioned above can often out-perform more complicated rules.

This is known technically as the "bias-variance dilemma," which arises in many areas of life. It can be written as an equation:

sity and Donald Sull from the Harvard Business School. Eisenhardt and Sull discovered that companies like eBay and Yahoo! which flourish in the turbulent and unpredictable markets of the modern economy, do so by following a small number of simple rules that allow them to respond with flexibility to fleeting opportunities.

The markets in which these companies function are on the edge of chaos, somewhat like the boids in Craig Reynolds' computer simulations. There is short-term order and organization, but the patterns fluctuate rapidly, and long-term trends are difficult or impossible to predict.

Why Do Simple Rules Work So Well in Resolving Complex Problems? (continued)

$$\text{total prediction error} = (\text{bias})^2 + \text{variance} + \text{noise}$$

Bias is just the difference between our picture of what is going on and the real picture, variance is a measure of the scatter in the data that we have, and noise is further scatter due to accidental effects.

As Gigerenzer and his colleague Henry Brighton explain in "Homo Heuristicus":

> To achieve low prediction error on a broad class of problems, a model must accommodate a rich class of patterns [i.e., it must incorporate many factors] in order to ensure low bias. . . . Diversity in the class of patterns that the model can accommodate is, however, likely to come at a price. The price is an increase in variance, as the model will have a greater flexibility, which will enable it to accommodate not only systematic patterns but also accidental patterns such as noise. When accidental patterns are used to make predictions, these predictions are likely to be inaccurate. This is why we are left with a dilemma: Combating high bias requires use of a rich class of models, while combating high variance requires placing restrictions on this class of models. We cannot remain agnostic and do both.

Companies like eBay, Yahoo! and Amazon exploit such markets by acting like hawks preying on the flock. They monitor the changing situation, probe for opportunities, build on successful forays, and shift among opportunities as circumstances dictate. Eisenhardt and Sull found that the rules fall into categories that we could find useful in preparing ourselves ahead of time for emergencies and rapid changes of circumstance. They list five such categories, which I briefly consider here:

1. "How-To" Rules

- The idea of a how-to rule is to specify in advance how we should act when a particular situation arises. Dell Inc., for example, has a rule for rapid reorganization stating that any business in the group must be split in two when its revenue hits $1 billion (it sometimes seems that a similar process happens in some marriages when the combined income reaches a certain figure!).

  It is worth considering some how-to rules that could be set in place in advance of particular events in our personal lives. An easily accessed list of emergency numbers provides one such rule (i.e., call the relevant numbers on the list in the event of an emergency). A box of relevant papers in case one partner dies is another.

2. Boundary Rules

- Cisco Systems, a provider of networking equipment, has a rule for its acquisitions-led business model that says it will acquire only companies with at most seventy-five employees, 75 percent of whom are engineers. Setting boundaries has certainly helped this company to avoid problems that result from an unbalanced workforce. It can equally help to avoid problems in our personal lives, whether the boundaries constrain children's play time or cap spending by either member of a couple without prior discussion.

3. Priority Rules

- Intel implemented a simple rule requiring manufacturing capacity to be based on a product's gross margin (the gross income divided by the net sales for that product—a measure of its profitability). It was by following this rule that the company was able to move from its core memory business to the highly profitable microprocessor niche.

  This has a strong parallel in our personal lives: allocate resources according to how important the activity or goods are to you. Some people might prefer to spend money on vacations. Others might prefer a new car. Assigning priorities is something we usually do automatically, but it is worth remembering that the allocation needs to be done.

  Those who don't plan allocations can soon find themselves in deep water. A friend of mine (a university professor) had the unpleasant experience of hearing a knock at the door and answering it to find an official representative of a credit card company on his doorstep. The representative asked to see my friend's credit card. When he produced it, the representative produced a large pair of scissors and cut the card in half before ceremoniously returning the pieces.

4. Timing Rules

- The global telecommunications company Nortel Networks relies on two simple timing rules: (1) project teams must always know when a new product has to be delivered to the leading customer to win an order, and (2) product development time must always be less than eighteen months. The first keeps the company in sync with leading-edge customers; the second forces the company to move quickly into new opportunities. The downside is that development teams must sometimes drop features of a product to keep to the development time.

Timing rules play out in our lives when we adopt different strategies for things we are doing only for ourselves and for stuff that involves others. When I am reading, for example, I make a great many notes, and I pursue ideas that I might like to write about in books. When I am writing a book, though, the deadline matters more than immediately satisfying my curiosity, and I must resist going off on interesting tangents, so I make a quick note to go back to them later instead.

5. Exit Rules

- The successful Danish hearing-aid company Orticon uses a simple exit rule to govern its many internal developmental projects: if a key team member moves from one project to another, the first project is simply shut down.

  Exit rules mean knowing when to cut your losses, something we can clearly plan for in our personal lives to avoid having to come up with some ad hoc strategy when faced with a decision. My own rule for cars, for example, is to keep going with the one that I have until the first big repair bill comes up and then change cars straight away.

In summary, simple heuristics have a lot to offer for dealing with decision making in complex situations. The best choice of heuristic depends on circumstances, and it is worth working out ahead of time just what the best choice will be when it comes to making a decision.

The alternative to simple heuristics is to search for patterns within the complexity. As I show in the next chapter, this, too, has its place when it comes to making the best decisions in complex situations.

NINE

Searching for Patterns: Imagination and Reality

If we can distinguish patterns within the depths of complexity, we may be able to use them as paths to guide us through the maze. There are two basic ways to do it: we can use our imaginations, or we can use statistics.

Imagination

In a famous episode of *Peanuts*, Lucy van Pelt, Charlie Brown, and Linus van Pelt use their imaginations to see patterns in clouds:

> *Lucy*: Aren't the clouds beautiful? They look like big balls of cotton. I could just lie here all day and watch them drift by. If you use your imagination, you can see lots of things in the cloud's formations. What do you think you see, Linus?
>
> *Linus*: Well, those clouds up there look to me look like the map of the British Honduras on the Caribbean. [*Pointing up*] That cloud up there looks a little like the profile of Thomas Eakins, the famous

painter and sculptor. And that group of clouds over there . . . [*Pointing*] . . . gives me the impression of the Stoning of Stephen. I can see the Apostle Paul standing there to one side.

Lucy: Uh huh. That's very good. What do you see in the clouds, Charlie Brown?

Charlie Brown: Well . . . I was going to say I saw a duckie and a horsie, but I changed my mind.

Linus and Charlie Brown's examples show that imagination can take many forms. Many people tend to think of it as the province of the arts and would be surprised to learn that imagination is equally important in science. It can be stimulated in both cases by events and patterns in the real world, but there is one vital difference: it is the job of the scientist to check the imagined pattern against the reality.

We begin our scientific careers as babies, when we are first confronted by a bewildering jumble of sensory experiences. Visual and aural sensations seem to be the most important, judging by the large areas of the brain that are used for the perception of visual and aural patterns.

Sounds gradually resolve themselves into the aural patterns of words and sentences, and sights become associated with the patterns of three-dimensional objects. The process of association is aided by the fact that the areas of the brain used for the perception of these patterns are also used for recalling patterns that have been perceived in the past.

Once children have established an association, they confirm it through experimental tests—touching the object, observing their parents' reactions when they use the word, stringing the words together in increasingly long sentences.

Imagination plays a vital role in all of this. It allows children to take shortcuts by providing a rapidly growing framework of relationships between sight and object, sound and meaning. It allows scientists to take similar shortcuts, only we don't call it "imagination"; we call it "hypothesizing." It's the same thing, though—dreaming up a picture of how things might be and then checking that picture against reality.

Many of those pictures have come from the perception of patterns lying deep in the mists of complexity. One of the best-known examples concerns the discovery of the periodic table by the Russian Dmitri Mendeleev.

Mendeleev was writing a textbook titled *Principles in Chemistry*, and he wanted to find some organizing principle on which he could base his discussion of the sixty-three chemical elements then known. He began by making cards for each of the known elements. On each card he recorded an element's atomic weight, valence, and other chemical and physical properties. Then he tried arranging the cards in various ways to see if any pattern emerged.

Eventually he was successful. When he arranged the elements in ascending order of their weights, he saw that their properties gradually changed. The metals lithium and beryllium were replaced by the nonmetals boron and carbon, and these were replaced in their turn by the gases nitrogen, oxygen, and fluorine. Suddenly, though, there was another metal—sodium. Mendeleev's first inspiration was to put the sodium card *under* the lithium one so as to start another row.

Soon he had a number of rows, and a problem. At several places there was a mismatch. Something didn't quite fit. Then came the second inspiration. By shifting the next cards one space to the right, everything fit again, except that there was now a gap between two of the elements (calcium and titanium).

Ultimately he found three gaps and had his third inspiration: that these would eventually be filled by elements not yet discovered. It wasn't long before the gaps were filled. Mendeleev published his book in 1869, and the three missing elements were discovered in

1875 (gallium), 1879 (scandium, which fit between calcium and titanium), and 1885 (germanium)—all with melting points and densities that he had been able to predict from the pattern of the table.

The periodic table is far from being the only discovery to emerge from a search for patterns and regularities in the depths of complexity. Many other famous discoveries have been made in a similar way. The physicist Murray Gell-Mann, for example, looked at the relationship between different components of the atomic nucleus and found that they fit a pattern of symmetry that could be explained only if there were eight basic units, seven of which were known and one of which remained to be discovered. It was, and it is, called the "omega-minus" particle. Gell-Mann was awarded a Nobel Prize for his insight.

Another Nobel laureate, biologist Jim Watson, used cut-out cardboard shapes to uncover the pattern that underlies the double helix of DNA. He cut them into the shapes of the four basic molecular components and then shifted them around on his desk until their paired patterns matched.

Watson justified his simple approach by saying that "nobody ever got anywhere by seeking out messes." In the early 1950s, when he made his discovery, he was surely right. Only since the 1980s have complexity scientists had the tools to tackle such messes in their totality.

When it comes to solving the problems of everyday life, the examples in this book have shown that in some cases we can use that totality in the form of swarm or group intelligence but in other cases we do best using simple rules that ignore the complexity and concentrate on one or two essential features of the problem. Can we do any better by using our imaginations to seek out patterns in everyday-life situations?

Mathematical biologist James Murray certainly succeeded when he noticed a correlation between the way in which newlyweds smiled when they talked to each other and the length of time that the marriage lasted. He collaborated with psychologists in a study of numerous couples that eventually produced a mathematical divorce formula. When the predictions of the formula were checked against reality, they proved to be correct a remarkable 94 percent of the time.

Murray used statistics to produce his formula and to verify its conclusions. Modern-day data miners do something similar when they use sophisticated software to ferret out trends, patterns, and correlations in the mass of information about us that is now held in various computer databases. But can we use statistics to help us tackle the complex problems of everyday life?

I believe that there are times when we can—not to ferret out patterns and trends but to help us distinguish illusory patterns from those that are real. There are two tools in particular that can help. One is Benford's law, which makes it easier for us to tell whether someone is trying to hornswoggle us with faked figures. The other is Ramsey's theorem, which helps us determine whether or not to believe the patterns we sometimes think we see.

NOTE: It requires elementary high school mathematics to follow some of the arguments in this chapter, but none at all to understand the conclusions, which I summarize at the end.

Benford's Law

Benford's law should really be called Newcomb's law, since it was originally discovered by the self-taught astronomer Simon Newcomb, and only rediscovered by the statistician Frank Benford, after whom it is named, some fifty-seven years later.

Newcomb had been appointed director of the U.S. Nautical Almanac Office in 1877 and had set himself the enormous task of recalculating all of the major astronomical constants. He used a telescope at the U.S. Naval Observatory in Washington, D.C., to recheck the positions of the stars and the planets and noticed that the book of logarithms provided for use by staff had very dirty pages at the front and comparatively clean pages at the rear.

Many of us have never seen such a book, but logarithms were an essential adjunct to fast calculation in the days before computers, since they allowed the operations of multiplication and division to be converted to the simpler and quicker operations of addition and subtraction. The figures were listed from lower to higher through the book, so the fact that the front pages of the book were dirtier meant that the lower figures were being used more often than the higher. But why?

Newcomb could think of only one explanation—that logarithms began more often with low digits like 1 or 2 than they did with higher digits like 7, 8, or 9. It seemed like a ridiculous assumption, but Newcomb discovered that there was a mathematical rationale that accounted for the distribution of the digits.

That reason is based on the recognition that when a set of numbers covers several orders of magnitude, it is their *logarithms* that are randomly distributed, rather than the numbers themselves. The deductive chain is rather esoteric, but it led Newcomb to a very simple formula for the distribution of first digits in any such set of numbers:

$$Fa = \log_{10}((a+1)/a)$$

where *Fa* is the frequency of the digit *a* in the first place of the numbers.

What this means is that in most lists of numbers from real life that cover a wide range, the digits in the first place should occur with the following probabilities:

Digit	Probability
1	30.1%
2	17.6%
3	12.5%
4	9.7%
5	7.9%
6	6.7%
7	5.8%
8	5.1%
9	4.6%

If they don't, you can bet that the list has been doctored.

Frank Benford found that the law applied to lists of street addresses, population statistics, death rates, stories in the *Readers' Digest*, and the drainage rates of different rivers. More recently Mark Nigrini, now a professor of accounting, realized while he was still a student that Benford's law could be used as a test for financial fraud, since fraudsters tend to randomize by giving equal weight to all digits. This has led to the discipline of forensic accounting, which tax officials now use routinely. One of its earliest applications by Nigrini was to President Bill Clinton's published tax accounts over thirteen years. They passed!

Many other applications of Benford's law are now being explored, from checking for falsification in clinical trials to the reality of projections in planning applications. All of these real-life applications arose just because Newcomb and Benford noticed patterns in a seemingly random mass of complex data, and had the courage to believe in the reality of what they saw.

Perceived patterns are not always real, however. The ancient Babylonians and Assyrians believed that there were relationships between events in the sky and events here on Earth and used the faulty logic of the post hoc fallacy to support their belief. If a lunar eclipse was followed by a good crop, for example, then to the Assyrian and

Babylonian mind it was obvious that the eclipse had caused the good crop. If the heliacal rising of Mars (when it is just above the Eastern horizon at dawn) coincided with the death of cattle from disease, then it was equally obvious that the rising of Mars had caused the disease. Once the faulty connection was established, it was enshrined in a book of omens. We may no longer believe in omens specifically, but many of us still think that events in the sky influence events here on Earth.

One such person is a friend of mine who holds a very high position in a prestigious international organization. He always carries with him a set of astrological charts, and he even claims that they predicted the time when he was held at gunpoint on the tarmac of a remote airfield.

When I asked why he didn't use the prediction to change his flight, he had no real answer. He did, however, have an answer to the criticism that astrology had never yet passed a scientific test of its validity. "Just look how everything links together," he said. "It's *obvious* that there is a pattern that links all of these things."

Unfortunately he is not a mathematician. If he were, I might have been able to convince him of just how wrong his argument is by introducing him to Ramsey's theorem.

Ramsey's Theorem

Frank Plumpton Ramsey was a Cambridge mathematician who applied his mathematics to philosophy, economics, and logic problems with equal dexterity. A tall, gangling hulk of a man, he lived for just twenty-six years before dying after an operation in 1929. By then, though, he had produced a plethora of ideas that are still deeply relevant to the problems of today.

One of those ideas concerned the proportion of its income that a nation should prudently save. The economist John Maynard Keynes called it "one of the most remarkable contributions to mathematical economics ever made" but went on to admit that "[Ramsey's] article is terribly difficult reading for an economist." It is pretty difficult reading for a noneconomist, too, as the following quotation—Ramsey's explication of his principle—shows: "The rate of saving multiplied by the marginal utility of money should always be equal to the amount by which the total net rate of enjoyment of utility falls short of the maximum possible rate of enjoyment."

Phew! In fact, what Ramsey had come up with was a complicated but mathematically rigorous way of calculating what proportion of our money we should put aside to cover eventualities. If modern politicians and business leaders had heeded his advice, the credit crunch might never have happened. Instead, we could be in the state of balance that Ramsey called "bliss."

Ramsey produced true bliss for mathematicians by providing them with a problem on which they are still working, and which is particularly relevant when it comes to finding real patterns in a complex mass of data. At its heart lies a revolutionary inspiration about connectivity and order in networks—of people, of events, and of ideas themselves.

The idea is now called Ramsey's theorem. Ramsey proved that as a group of objects gets larger and larger, patterns must inevitably emerge within the group. If we are in a group of six people at a party, for example, there must inevitably be at least three of us who are linked through knowing each other, or three who are linked in the negative sense that they do *not* know each other.

This may be hard to believe, but we can easily prove it by common sense in this simple case. Select any one of six guests at a party. He either knows or does not know three of the others.

If he knows three others, and any two of these three know each other, then he joins this pair in an acquaintance triangle. Otherwise the three others are mutual strangers.

If he does *not* know three others, we can apply a mirror image of the same logic. If any two of these three do not know each other, that creates a triangle of mutual strangers. Otherwise, the three others are mutual acquaintances.

This may sound rather long-winded, but at least it is understandable. For larger numbers, the problem itself still sounds simple, but the reasoning gets so complicated that even the best mathematicians are still struggling with it. It is a part of graph theory (concerned with the linkages between points in a network), which has many real-world applications, but where progress has been slow and sometimes excruciatingly painful. It was not until 1955, for example, that mathematical techniques became sufficiently advanced to work out the answer for a party where at least four people know each other, or there are four who are mutual strangers (the answer turned out to be eighteen). When it gets to five either way, mathematicians still don't have a definite answer; all that they know is that the number must be between 43 and 49.

These numbers (the minimum number to be sure of each condition) are called Ramsey numbers. They are written as $R(m,n)$, which is the number of people necessary for there to be either m mutual acquaintances or n mutual strangers. So $R(3,3) = 6$; $R(4,4) = 18$; and $R(5,5)$ lies somewhere between 43 and 49.

We can also have asymmetric Ramsey numbers. $R(4,5)$, for example, is the number needed to guarantee four mutual strangers and five mutual acquaintances (or vice versa). Its value is 25.

The larger Ramsey numbers are ferociously difficult to calculate, but mathematicians have at least succeeded in calculating bounds for many for them. $R(6,6)$, for example, lies somewhere between 102 and 165, and $R(7,7)$ is somewhere between 205 and 540. $R(19,19)$ lies somewhere between 17,885 and 9,075,135,299!

The significance of Ramsey's theorem for the real world is that it lets us know whether perceived chains of connection are likely to be statistical artifacts or whether they are likely to represent a real underlying pattern. It all depends on the size of the group. For example, we can *always* establish a chain 7 links long somewhere within a group of 540 people or events, but the mere existence of such a chain in this case does not prove that it has any underlying significance. If we find a chain 7 links long in a smaller group—100, say—then the existence of the chain *does* provide a priori support for the existence of an underlying cause.

When we see a chain that connects points or events within a huge mass of data (such as that found in an astrological chart or in global weather patterns), Ramsey's theorem tells us that we should look for other evidence of an underlying cause before concluding that the chain is meaningful. In the case of global weather patterns that suggest global warming, that other evidence has been forthcoming in the form of measurable temperature and rainfall changes and the rapid melting of the polar icecaps. In the case of astrology, there is no such hard evidence—just chains of connection that Ramsey's theorem tells us could well be statistical artifacts that are no more than a matter of chance.

Synopsis

We can usefully add imagination and the perception of patterns to our kit of tools for tackling the complex problems of everyday life. The main requirement is that we should be able to check the perceived pattern against reality, which we can do by the scientist's method of doing an experiment or by using statistics such as that encapsulated in Benford's law and Ramsey's theorem.

Benford's law tells us that the first digits in a list of numbers that covers several orders of magnitude should be distributed according to the table on page 161. If they aren't, chances are that the list has been doctored.

Ramsey's theorem tells us that there will *always* be chains of connection in any sufficiently large group of facts, people, or events. It also identifies group sizes above which such chains are likely to be statistical artifacts without any deeper meaning.

Together with the uses of swarm and group intelligence that are discussed in the early chapters, and the simple rules that are listed in the middle chapters, these mathematically-based laws complete our kit of tools for tackling everyday complexity. In the next chapter I summarize just what those tools can do for us, and how we can best use them.

There is just one proviso. We live in a complex world, and its emergent behaviors mean that we cannot always predict how situations may end up. Simple rules, patterns, and formulae can often help us steer our way through, but in the end it is the complexity that rules. OK?

TEN

Simple Rules for Complex Situations

Complexity science tells us that simple interactions between neighbors can lead to complex group behaviors like swarm intelligence, the whole becoming greater than the sum of its parts. Sometimes we can identify and take advantage of the rules that gave rise to the complexity. On other occasions we can take advantage of the complexity itself by using group or swarm intelligence to help resolve problems. We can also use simple heuristics, consensus techniques, networking, data mining, pattern recognition, and even mathematical theorems to help guide us through the complexities of life.

When we are faced with complex situations, we should look back at the science of complexity and remember these top ten tips.

1. Develop the virtues of swarm intelligence in family, community, and business environments by providing a platform that encourages people to see themselves as stakeholders rather than shareholders. You can also develop swarm intelligence in your group by using cell phones to turn yourselves into a smart mob.

2. Lead from the inside (if possible with a coterie of like-minded friends or colleagues), but take care not to let other members of the group know what you are doing. Just head in the direction that you want to go, and leave it to the laws of the swarm to do the rest.
3. When networking, find, use, or establish those few long-range links that bring clusters together into a small world with only a few degrees of separation.
4. If you are with a crowd in a dangerous situation, use a mixed strategy for escape; follow the crowd 60 percent of the time, and spend the other 40 percent searching out escape routes on your own.
5. If you want to give yourself the best chance of choosing the very best option in a situation doesn't allow you to go back to the options that you have rejected, look at 37 percent of those available, then choose the next one that is better than any of them. This will give you a 1 in 3 chance of finding the best option, and a very high chance of finding one in the top few percent. You can also improve your chances by lowering your aspirations slightly. If you want an alternative that is in the top 10 percent of all alternatives, look at 14 percent of those on offer, then choose the next one that is better than any that you have already seen.
6. If you want to persuade a large group of people, or even start a craze, don't rely on persuading someone with influence to pass the message on. It is far better to try for a critical mass of early adopters—people who will take the idea or product up after a single exposure.
7. When confronted with a mass of data that you need to use as the basis for a decision, first use Benford's law to check that the data haven't been faked.

8. Don't believe that chains of connection have an underlying meaning unless you have some reason other than the connections themselves.
9. Pedestrian crowds tend to self-organize into flowing rivers. Follow the rivers; don't get stuck on the banks.
10. When planning a complicated road trip across a city, build in as many right turns as possible (or left turns if you happen to be in a country with left-hand drive).

Here are the other rules that have emerged from our study of complexity in real life:

Rules from Social and Behavioral Studies

- When taking an unofficial shortcut across a field or park, don't just blindly follow the tracks of those who have gone before. Work out the shortest route for yourself. Experiments have shown that it will often be shorter. Start a new trend!
- If there is a warning of danger, believe it, and act on it promptly. Don't hang around waiting for confirmation.
- If you are on the outside of a dense crowd, back off, and try to persuade others to do the same.
- If you are given a pair of alternatives to choose from, and you recognize only one, then (in the absence of other information) choose the one that you recognize.
- If recognition is all that you have to go on, and you recognize more than one alternative, go with the one that you recognized most easily.
- Look for cues that might help you to make a choice between options, and choose the option that has the greatest excess of positive over negative cues.

- Alternatively, when faced with a choice between two options, look for cues and work through them in order of your expectation that they will lead to the best choice. Make your choice as soon as you find a cue that distinguishes between the alternatives.
- If there is a default option to do nothing, then often it is better to take this option.
- As an easy option, take the course of action that the majority of your peer group is taking.
- Follow the example of those who have succeeded, but make sure that you have the same qualities that they do!

Rules from Nature

- If your only clue in a situation is how others are behaving, use the quorum response, which will make your likelihood of choosing the best option increase steeply with the number of people already committed to that option. If you can supplement the quorum response with additional information, this will help your chances.

Rules from Business

- Avoid the perils of groupthink by escaping temporarily from the group environment, doing some independent thinking, and committing yourself to the conclusions of that thinking before returning to the group.
- Plan ahead for emergency situations and others in which you need to make rapid decisions by deciding in advance on how-to, boundary, priority, timing, and exit rules.
- Spread your bets evenly. Instead of choosing one alternative over another, allocate your resources equally to each.

Rules from Computer Modeling

- In working your way through a crowd with a group of friends, stick together closely, but also try to get as many strangers as possible to stick with you.
- When trying to bring an issue to the notice of a group or the public as a whole, don't be a one hit wonder; plan to bring different aspects to the fore over time.
- When performing any task, keep an eye on those nearby who are performing similar tasks. If they are performing better than you are, don't be too proud to copy them!
- Win-stay, lose-shift. In recurring situations that require cooperation, always cooperate at first, but then do in the second encounter whatever the other person did in the previous encounter (i.e., cooperate or don't cooperate).

Rules from Mathematics

- When tackling state estimation problems that involve figuring out the value of something, use the cognitive diversity of the group. Just remember that the individuals in the group need to come to independent judgments before pooling their conclusions to take an average.
- With problems for which there is one definite correct answer, the majority verdict of the group provides the best chance of finding that answer. The bigger the group, the better the chance of the majority being right.
- To achieve consensus by voting, choose the voting method that is most practicable, rather than the one closest to ideal. There is no ideal voting method, and there never can be.
- If you want to establish a chain of contacts between yourself and someone you don't know personally, provide an incentive

that can be passed along so that others will keep the chain going. Also, choose your first contact to be someone who is closest in social distance to the target.

- When networking, choose and use hubs.
- If you are looking at a series of alternatives, and cannot return to one once you have rejected it, then satisfice by choosing the first one that exceeds your aspiration level.

One final message: use the rules wisely. Remember that life is complex and that emergent patterns can't always be predicted from simple rules, even though such rules lead to them. By using the rules we have uncovered in this book, though, you will certainly give yourself the best chance of emerging unscathed from the mire of complexity that so often seems to have us trapped.

NOTES

Introduction

1 ***ninety-seven locusts sat down to watch [Star Wars]*** F. Claire Rind and Peter J. Simmons, "Orthopteran DCMD Neuron: A Reevaluation of Responses to Moving Objects. I. Selective Responses to Approaching Objects," *Journal of Neurophysiology* 68 (1992): 1,654–1,666.

Researcher Claire Rind was "80% chuffed, 20% mortified" when she and her collaborator, Peter Simmons, were awarded an Ig Nobel Prize for their experiments (BBC News Report, March 13, 2006, http://news.bbc.co.uk/2/hi/uk_news/magazine/4801670.stm). It wasn't even a prize for science; it was for *peace*! Harvard University's Marc Abrahams, who organizes the prizes, explained the reason when I asked him why he placed this project in the peace category: "Locusts are generally associated with pestilence and widespread disaster. Rind and Simmons used a movie about cosmic war to gain insight on how these much-maligned creatures go about avoiding even so small a conflict as a collision with another locust."

For details of the Ig Nobel prizes, see improbable.com. Speaking as a former recipient, I can understand how Claire Rind felt. At least my prize was for a deliberately lighthearted project (using physics to work out the best way to dunk a cookie) that was intended to bring public attention to science, and not for a serious piece of science.

1 ***design of collision avoidance systems for cars*** Richard Stafford, Roger D. Santer, and F. Claire Rind, "A Bio-Inspired

Visual Collision Detection Mechanism for Cars: Combining Insect Inspired Neurons to Create a Robust System," *BioSystems* 87 (2007): 164–171.

See also http://www.staff.ncl.ac.uk/claire.rind/try1.htm.

2 ***superorganism*** The word was coined by the American entomologist William Morton Wheeler in 1928 (*Emergent Evolution and the Development of Societies* [New York: W. W. Norton, 1928], following Herbert Spencer's use of the term *super-organic* in *The Principles of Sociology* (London: Williams and Norgate, 1874–1875). Spencer has sometimes been described as the originator of the concept, but he was actually making a distinction between the organic and the social, rather than assigning an overall identity. In any case, there are many earlier claimants to the general idea that societies have the attributes of organisms (e.g., Alfred E. Emerson, "Social Coordination and the Superorganism," *American Midland Naturalist* 21 [1939]: 182–209).

2 ***insects such as locusts, bees, and ants*** Entomologists call the latter two "social insects" because they display reproductive division of labor, which locusts do not, even though they display highly social behavior when they swarm.

2 ***self-organization*** For some nice examples of self-organization in society, see Dirk Helbing, ed., *Managing Complexity: Insights, Concepts, Applications* (Berlin: Springer, 2008).

2 ***atomic diameters*** The diameter of an atom is of the order of one nanometer (one billionth of a meter).

2 ***swarm intelligence*** Most scientists would agree with the opinion of authors Hongbo Liu, Ajith Abraham, and Maurice Clerc (from China, South Korea, and France respectively) that swarm intelligence emerges from "a chaotic balance between individuality and sociality" ("Chaotic Dynamic Characteristics in Swarm Intelligence," *Applied Soft Computing* 7 [2007]: 1,019–1,026).

3 ***the edge of chaos*** This well-known phrase was coined by Los Alamos computer scientist Chris Langton in 1990 to describe the circumstances under which the capacity to compute might emerge spontaneously in physical systems (including living ones) as a part of their self-organization.

("Computation at the Edge of Chaos: Phase Transitions and Emergent Computation," in *Proceedings of the Ninth Annual International Conference of the Center for Nonlinear Studies on Self-organizing, Collective, and Cooperative Phenomena in Natural and Artificial Computing Networks* [Amsterdam: North-Holland, 1990], 12–37.)

The applicability of the phrase to living systems was popularized by Roger Lewin in his 2000 book *Complexity: Life at the Edge of Chaos* (Chicago: University of Chicago Press).

Elizabeth McMillan provides an excellent history of the historic relationship between chaos theory and complexity science in *Complexity, Organizations and Change: An Essential Introduction* (New York: Routledge, 2004).

3 ***Rayleigh-Bénard cells*** There is a good description and an image at http://www.etl.noaa.gov/about/eo/science/convection/RBCells.html.

4 ***complex adaptive systems*** For a more complete technical summary, see John H. Miller and Scott E. Page, *Complex Adaptive Systems: An Introduction to Computational Models of Social Life* (Princeton: Princeton University Press, 2007).

Chapter 1

9 ***swarm intelligence*** This area of research is now so important that there is a journal of the same name (published by Springer).

9 ***swarms . . . schools . . . flocks*** There are many evocative collective nouns for animal groupings. Some of my favorites among the less well known are an exaltation of larks, an unkindness of ravens, a crash of rhinoceroses, and (best of all) a cry of actors. Other collective nouns that have been suggested, but which I hope are not exemplified by this book, include a pretension of intellects and a pomposity of professors.

9 ***extrasensory perception*** E. Selous, *Thought Transference (or What?) in Birds* (London: Constable, 1931), 139.

9 ***sacrifice its individuality and become a puppet*** If this were to apply to humans, the resultant society would be like that described by Aldous Huxley in *Brave New*

World (London: Chatto & Windus, 1932), in which groups of individuals were psychologically conditioned as infants to behave identically and to accept their role in society. Fortunately, there is no need for such a dramatic sacrifice. Studies of swarm intelligence have shown that the key is not losing our individuality but learning how to interact appropriately with those who are closest to us. If we do it the right way, swarm intelligence emerges as a natural consequence.

9 ***[swarm behavior] emerges naturally from simple rules of interaction*** For an excellent summary, see Simon Garnier, Jacques Gautrais, and Guy Theraulaz, "The Biological Principles of Swarm Intelligence," *Swarm Intelligence* 1 (2007): 3–31.

10 ***plays a key role in our own society*** The self-organization that permits the swarm to form and guides its actions is equally apparent in the functioning of our brains, the performance of our immune systems, the organization of our society, and the global ecological balance. In all of these cases, local interactions—between molecules, living cells, people, and mixtures of species respectively—produce large-scale complexity.

The above list of examples was used by pioneering complexity scientist John Holland to introduce his wonderfully prescient 1994 lecture "Complexity Made Simple" at the now-famous Santa Fe Institute in New Mexico. It was the first in an annual series of Stanislaw M. Ulam memorial lectures, and it laid out the plan for future work on complexity at that institute and elsewhere. The lecture was aimed at a general, science-interested audience, and it became the basis for Holland's book *Hidden Order: How Adaptation Builds Complexity* (New York: Basic Books, 1996).

As Holland points out, complexity does not mean chaos. It means the formation of patterns whose component parts are intricately interconnected and interdependent. The patterns may be stable over a long period of time, stable over shorter periods of time, or relatively transient. Long-term stability is to be found in crystals, seashells, living cells, civilizations, and galaxies. Some examples of shorter-term stability but long-term instability are political alliances;

networks of suppliers, manufacturers, and retailers; some ecosystems; and quite a few marriages. Transient self-organized patterns arise in rafts of bubbles in a bath, lines of shoppers in a supermarket, doing the wave in a stadium, the spinning vortex of a tornado, and of course swarms, flocks, shoals, and herds of animals, not to mention human crowds.

10 ***companies run by swarm intelligence*** Peter Gloor and Scott Cooper, "The New Principles of a Swarm Business," *MIT Sloan Management Review* (spring 2007): 81–84. See page 101 for a detailed discussion.

10 ***radical approach to problem solving*** The approach is called "particle swarm optimization" (page 43). For its remarkable range of applications, see Ricardo Poli, "Analysis of the Publications on the Applications of Particle Swarm Optimisation," *Journal of Artificial Evolution and Applications* (2008), http://www.hindawi.com/journals/jaea/2008/685175.html.

10 ***Swarmfest*** Swarmfest is an annual conference organized by a volunteer group of scientists who are interested in the computer simulation of swarm behavior, the Swarm Development Group (swarm.org). The group was founded in September 1999.

11 ***Martin Lindauer was caught in a particularly bizarre situation*** M. Lindauer, "Schwarmbienen auf Wohnungssuche," *Zeitschrift für vergleichende Physiologie* 37 (1955): 263–324.

Lindauer died in 2008. His obituary appeared in *Nature* (456 [2008]: 718). In a commentary on the obituary (*Nature* 457 [2009]: 379), William Abler, from the Chicago Field Museum, pointed out that Lindauer produced "the best-ever experimental evidence for evolution" in a beautifully simple experiment. "Living in an enclosed, sheltered space," says Abler "the honeybee *Apis mellifera* performs its communicative dance on a vertical surface in the dark, using gravity as a substitute for the direction of the Sun. By depriving them of a vertical surface and giving them a direct view of the Sun, Lindauer forced them to revert to the more primitive, Sun-directed dance of their dwarf Indian relative, *Apis florae.*"

11 ***two Brazilian scientists decided to follow schools of piranhas*** Ivan Sazimat and Francisco A. Machado, "Underwater Observations of Piranhas in Western Brazil," *Environmental Biology of Fishes* 28 (1990): 17–31.

11 ***Aristotle . . . thrust his bearded visage under the waters*** Jason A. Tipton, "Aristotle's Observations of the Foraging Interactions of the Red Mullet (*Mullidae: Mullus spp*) and Sea Bream (*Sparidae: Diplodus spp*)," *Archives of Natural History* 35 (2008): 164–171.

12 ***Brian Partridge was studying Atlantic pollock*** Brian L. Partridge, "Internal Dynamics and the Interrelations of Fish in Schools," *Journal of Physiology* 144 (1981): 313–325.

14 ***logistic difference equation*** R. M. May, "Simple Mathematical Models with Very Complicated Dynamics," *Nature* 261 (1976): 459–467.

14 ***populations would initially grow*** A population grows exponentially when it increases by the same percentage of its previous value over each successive (equal) period of time.

14 ***scientific intelligence expert R. V. Jones*** Jones was the recipient of an award from the CIA (appropriately known as the R. V. Jones Intelligence Award) for "scientific acumen applied with art in the cause of freedom."

15 ***a machine that could laugh by itself*** Described in R. V. Jones, *The Wizard War: British Scientific Intelligence 1939–1945* (New York: Coward, McCann & Geoghegan, 1978). The title in the United Kingdom is *Most Secret War* (the subtitle is the same).

Perhaps there will never be a machine that can laugh by itself, but see Isaac Asimov's short story "Jokester" for a computer that is capable of analyzing the sense of humor.

15 ***the collapse of Washington Mutual*** Eric Dash and Andrew Ross Sorkin, "Government Seizes WaMu and Sells Some Assets," *New York Times*, September 26, 2008, A1.

15 ***"Ford effect"*** The car-buying illustration is due to economist W. Brian Arthur, a founding figure of the Santa Fe Institute, although he did not name specific cars in the original example (M. Mitchell Waldrop, *Complexity: The Emerging Science at the Edge of Order and Chaos* [New York: Simon & Schuster, 1992], 45).

Early on in the development of complexity theory, Arthur and his Soviet colleagues Yuri Ermoliev and Yuri Kaniovski showed that the Ford effect can produce not only a strong bias toward one product or the other but also any ultimate ratio of sales at all, depending on tiny differences in the initial conditions (W. Brian Arthur, "Competing Technologies, Increasing Returns, and Lock-In by Historical Events," *The Economic Journal* 99 [1989]: 116–131). The practical consequence of this is that the free market (and individual freedom with it) "might *not* produce the best of all possible worlds" (Waldrop, *Complexity*, 48).

16 ***"The Day the Dam Broke"*** First published on July 29, 1933, in the *New Yorker* this story is available in James Thurber, *My Life and Hard Times* (New York: Perennial Classics, 1999).

16 ***chain reaction*** For a nice visual example of a nuclear chain reaction, see http://www.lon-capa.org/~mmp/applist/chain/chain.htm.

It is worth pointing out that chain reactions are not always bad. Our ability to make multiple copies of DNA is an essential process, for example, that depends on a biological chain reaction called the "polymerase chain reaction."

17 ***Negative feedback is frequently used to "correct" errors*** The use of feedback processes is the subject of a huge literature. Here I have only introduced the basic ideas without mentioning the complications. One complication in the case of negative feedback is the possibility of overcorrection, which can produce a swinging instability that is every bit as severe as the runaway effects produced by positive feedback.

17 ***invisible hand*** Adam Smith first used the term "invisible hand" in *An Inquiry into the Nature and Causes of the Wealth of Nations* Book 4, Chapter 1 (Edinburgh, 1776).

17 ***a total of eight criteria for collective adaptability*** John H. Miller and Scott E. Page, *Complex Adaptive Systems: An Introduction to Computational Models of Social Life* (Princeton: Princeton University Press, 2007), 93–101.

Other workers have produced lists of more-or-less equivalent criteria. The first, which has served as a basis for the development of the whole field, was by John Holland in his

1994 lecture "Complexity Made Simple" (see *Hidden Order: How Adaptation Builds Complexity,* New York: Basic Books, 1996). In addition to nonlinear interactions between agents, Holland says that the group as a whole needs:

- Aggregation (individual agents need some way to remain linked together)
- Flow (Something must pass from agent to agent within the group. This may be information, or it may be something more material) and
- Diversity (not an essential requirement, but often a very useful one).
- Individual agents also need to have several specific abilities:
- They should be able to *recognize* other individuals and respond to them.
- They should in some sense be able to *predict* what the effect of a particular action might be. In the case of a bacterium the "prediction" may be no more than the built-in behaviour that would correspond to "if I swim up this chemical gradient, then I will find food." In the case of a fish in a school, it might correspond to the notion that swimming sideways would not be a good idea, because this might lead to a collision with a neighbor.
- Finally, they should behave as if they have some sort of internal picture (either learned or hardwired) that lets them relate their sensory experience to previous experience, and that they can use to guide them when related experiences happen.

Animal behavior expert David Sumpter's proposed list is slightly different, but it is particularly relevant to the consensus behavior of social animals (D. J. T. Sumpter, "The Principles of Collective Animal Behaviour," *Philosophical Transactions of the Royal Society* 316 [2006]: 5–22).

Sumpter's criteria (which I slightly paraphrase here) are:

- *Individual variability* If all of the animals in a group respond in the same way, this can sometimes be an advantage, but could also lead to disaster. For example, if every bee always collected the same type of food from the same place then their colony's [range of] nutritional needs could not be met.

- *Response thresholds* Behaviors in response to a stimulus may need to change above a certain threshold, in the way that bumblebees switch to fanning behavior when the temperature of the hive gets too high.
- *Redundancy* Unlike computers, "insect societies never crash." This is partly because they are made up of vast numbers of replaceable units, so that if a few individuals die or are removed, their function is not completely lost to the group.
- *Synchronization* Synchrony has its advantages, as marching locusts and human armies know. When individuals get out of step, negative feedback helps to get them back in. Some ant colonies synchronize bouts of resting and activity. If their efficiency during active phases is enhanced by positive feedback, then there is a net gain for the group (B. J. Cole "Short-term activity cycles in ants: generation of periodicity through worker interaction," *American Naturalist* 137 [1991]: 244–259).
- *Selfishness* The ultimate advantage of "intelligent swarming" and other complex behaviors is that every individual gets a benefit over that that they would get by acting alone. But game theory proposes this as a puzzle: why not cheat and get a greater benefit still? The logic works until all members of the group use it, and all lose out. *Somehow* the rules of individual interaction that lead to beneficial complex group behaviors must break this logical deadlock. It is not yet clear just how they do it.

19 ***virtual individuals are given specific rules of behavior*** An example of this sort of approach is the self-propelled particle (SPP) model. See C. Tamás Vicsek et al., "Novel Type of Phase Transition in a System of Self-Driven Particles," *Physical Review Letters* 75 (1995): 1,226–1,229.

20 ***mimic the way in which social animals . . . use swarm intelligence*** The early history of the approach to problem solving via computerized swarm intelligence (called "ant

colony optimization") is described by Eric Bonabeau (a member of the Santa Fe Institute) and his colleagues M. Dorigo and G. Theraulaz in "Inspiration for Optimization from Social Insect Behaviour," *Nature* 406 (2000): 39–42.

21 ***the logistic difference equation*** Originally proposed by P. F. Verhulst in 1838 ("Notice sur la loi que la population pursuit dans son accroissement," *Correspondence Mathematique et Physique* 10 [1838]: 113–121). For an explanation with plentiful graphs, see http://mathworld.wolfram.com/LogisticEquation.html.

Chapter 2

23 ***locusts*** For an informative summary, see Stephen J. Simpson and Gregory A. Sword, "Locusts," *Current Biology* 18 (2008): R364–R366.

23 ***stimulates them to produce . . . serotonin*** This is true for the African desert locust, but serotonin is now known *not* to be involved in the similar behavior of the Australian plague locust (Greg Sword, personal communication, 2009). So there appears to be a convergent evolution of behavioral changes in different locust species, with different underlying neurophysiological mechanisms.

In the African desert locust a triggering factor for the production of serotonin is physical contact with other locusts, which can be simulated in the laboratory by tickling the back legs with a paintbrush. The sight and smell of other locusts can also act as a trigger for the manufacture of the chemical neurotransmitter (Michael L. Anstey et al., "Serotonin Mediates Behavioral Gregarization Underlying Swarm Formation in Desert Locusts," *Science* 323 [2009]: 627–630; P. A. Stevenson, "The Key to Pandora's Box," *Science* 323 [2009]: 594–595).

23 ***a hundred billion locusts*** For more detail and pictures, see http://animals.nationalgeographic.com/animals/bugs/locust.html.

Mark Twain once noted that "nature makes the locust with an appetite for crops; man would have made him with an appetite for sand" (*Following the Equator: A Journey Round the*

World [Hartford: American Publishing Co., 1897; Washington, DC: National Geographic Society, 2005], http://www.literaturecollection.com/a/twain/following-equator/31/).

24 ***modern plagues affect the livelihoods of 10 percent of the world's population*** Stephen J. Simpson and Gregory A. Sword, "Locusts," *Current Biology* 18 (2008): R364.

24 ***a dramatic and rapid transition occurs*** J. Buhl et al., "From Disorder to Order in Marching Locusts," *Science* 312 (2006): 1,402–1,406. To observe this behavior, the researchers placed the locusts in a "Mexican hat" arena and used sophisticated tracking software to monitor the position of each locust.

The first example of such a transition was found in ants (M. Beekman, D. J. T. Sumpter, and F. L. W. Ratnieks, "Phase Transitions between Disorganised and Organised Foraging in Pharoah's Ants," *Proceedings of the National Academy of Sciences of the United States of America* 98 [2001]: 9,703–9,706).

24 ***somewhat disordered movement . . . changes to highly aligned marching*** The Irish comedian Dave Allen produced a wonderful example in a skit. People emerged from successive doorways and joined a tightly packed, synchronously walking, single file line of people who continued to march along until they reached their destination—a sardine packing factory.

24 ***college students on a campus and children on a playground*** L. F. Henderson, "The Statistics of Crowd Fluids," *Nature* 229 (1971): 381–383.

24 ***Video studies of pedestrians*** Dirk Helbing, Peter Molnar, Illes J. Farkas, and Peter Setany, "Self-Organizing Pedestrian Movement," *Environment and Planning B: Planning and Design* 28 (2001): 361–383; Kang Hoon Lee, Myung Geol Choi, Qyoun Hong, and Jehee Lee, "Group Behavior from Video: A Data-Driven Approach to Crowd Simulation," in *Proceedings of the 2007 ACM SIGGRAPH/Eurographics Symposium on Computer Animation*, edited by D. Metaxas and J. Popovic (2007), 109–118, http://portal.acm.org/citation.cfm?id=1272690.1272706.

25 ***the simple desire not to be eaten by the locust behind*** See Sepidah Bazazi et al., "Collective Motion and Cannibalism in Locust Migratory Bands," *Current Biology* 18 (2008): 735–739.

25 ***keep your distance*** This rule produces many spectacular self-organized patterns in nature. Opals, for example, are composed of small silica spheres that have formed a semiordered array under the influence of a mutual keep-your-distance rule (Pamela C. Ohara et al., "Crystallization of Opals from Polydisperse Nanoparticles," *Physical Review Letters* 75 [1995]: 3,466–3,470). The resultant order is sufficient to scatter white light into its myriad colors in a manner similar to the way the ordered array of grooves on a compact disk does when it is held up to the light, as I show in my book *Weighing the Soul: Scientific Discovery from the Brilliant to the Bizarre* (New York: Arcade, 2004), 58.

25 ***Craig Reynolds*** Reynolds worked for Symbolics Inc. at the time when he created boids. He now works for Sony Computer Entertainment.

25 ***original animation is still worth a look*** Reynolds' original computer animation can be found at http://www.red3d.com/cwr/boids/. A more recent example, which includes a beautiful Australian bush background, can be found at http://www.vergenet.net/~conrad/boids/. Note that in the latter case the programmer used the equivalent rule 3 (boids try to match velocity with near boids) explicitly.

26 ***conference on "artificial life"*** M. Mitchell Waldrop, *Complexity: The Emerging Science at the Edge of Order and Chaos* (New York: Simon & Schuster, 1992), 235–240.

26 ***one boid crashed into the pole*** M. Mitchell Waldrop, *Complexity: The Emerging Science at the Edge of Order and Chaos* (New York: Simon & Schuster, 1992), 242.

28 ***Reynolds' three rules*** Reynolds' rules have been used to model the behavior of schools of fish. The patterns that emerge closely resemble those found in real schools, including rather oblong shapes and a high frontal density, which are supposed to help the shoal protect itself from predators.

The models have nevertheless been criticized because some of their assumptions—such as that the fish swim at a constant average speed—are convenient for the computer programmer but may not be so convenient for the fish (J. Parrish and S. V. Viscido, "Traffic Rules of Fish Schools: A Review of Agent-Based Approaches," in *Self-Organisation and the Evolution of Social Behaviour*, edited by Charlotte Hemelrijk (Cambridge: Cambridge University Press, 2005). Another unrealistic assumption is that the number of nearby fish that any particular fish can see and take notice of is always the same. In reality, the more dense the school, the fewer visible neighboring fish.

One recent model that has taken account of these and other real-life factors predicts similar shoal shapes, but for different reasons (Charlotte K. Hemelrijk and Hanno Hildenbrandt, "Self-Organized Shape and Frontal Density of Fish Schools," *Ethology* 114 [2008]: 245–254). The high density of fish at the front was formerly thought to be caused by a sort of traffic jam as fast fish from behind pushed up. It is now thought to happen because the fish at the front slow down and fall back to the shoal.

27 ***taken up enthusiastically by the computer animation industry*** Its first application was in the 1992 Tim Burton film *Batman Returns*, in which it produced simulated bat swarms and penguin flocks. It continues to be used in films and video games as the basis for evermore sophisticated animations. Reynolds himself has produced many such animations for Sony's PlayStation.

27 ***[Reynolds' model] is still used today*** See for example Craig Reynolds, "Big Fast Crowds on PS3," in *Proceedings of the 2006 ACM SIGGRAPH Symposium on Videogames* (2006): 113–121, http://portal.acm.org/citation.cfm?id=1183333.

27 ***computer simulations*** J. Buhl et al., "From Disorder to Order in Marching Locusts," *Science* 312 (2006): 1,402–1,406.

27 ***self-propelled particle*** In this approach the "particles" interact through velocity-dependent forces. It has been developed in considerable detail since its original introduction by

Tamás Vicsek et al. ("Novel Type of Phase Transition in a System of Self-Driven Particles," *Physical Review Letters* 75 [1995]: 1,226–1,229). There have been mathematics-based demonstrations that both symmetric and asymmetric attraction/repulsion laws can produce stable swarms (V. Gazi and K. M. Passino, "Stability Analysis of Swarms," *IEEE Transactions on Automatic Control* 48 [2003]: 692–697; Tianguang Chu, Long Wang, and Shumei Mu, "Collective Behavior Analysis of an Anisotropic Swarm Model," *Proceedings of the 16th International Symposium on Mathematical Theory of Networks and Systems [MTNS 2004]*, edited by B. de Moor et al. [2004]).

One recent use has been to investigate how living cells can aggregate to form organized structures (Julio M. Belmonte, "Self-Propelled Particle Model for Cell-Sorting Phenomena," *Physical Review Letters* 100 [2008]), something with which I was intimately involved at one part of my research career when I collaborated in experiments designed to understand the forces involved in *Dictyostelium* slug movement (P. H. Vardy et al., "Traction Proteins in the Extracellular Matrix of *Dictyostelium Discoideum* Slugs," *Nature* 320 [1986]: 526–529).

28 ***early* Star Wars *experiments*** These experiments were described in the introduction. See F. Claire Rind and Peter J. Simmons, "Orthopteran DCMD Neuron: A Reevaluation of Responses to Moving Objects. I. Selective Responses to Approaching Objects," *Journal of Neurophysiology* 68 (1992): 1,654–1,666.

28 ***Later experiments, in which the locusts were allowed to fly freely*** Roger D. Santer, Peter J. Simmons, and F. Claire Rind, "Gliding Behaviour Elicited by Lateral Looming Stimuli in flying Locusts," *Journal of Comparative Physiology* 191 (2005): 61–73.

In these experiments, the spaceships were replaced by computer-generated circular disks that expanded and then shrank so that they appeared to move toward the locust and then move away, mimicking the approach and retreat of predators or other locusts in the swarm.

29 ***waggle dance*** This was first analyzed by the German ethol-

ogist Karl von Frisch, who received a Nobel Prize for his painstaking work ("Decoding the Language of the Bee," Nobel Lecture, December 12, 1973, http://nobelprize.org/nobel_prizes/medicine/laureates/1973/frisch-lecture.html). There were some subsequent doubts about the accuracy of his picture, but it has been vindicated by the recent work of Joseph Riley and his collaborators (J. R. Riley et al., "The Flight Paths of Honeybees Recruited by the Waggle Dance," *Nature* 435 [2005]: 205–207).

Frisch's original work was concerned with the way in which bees use the waggle dance to guide other individuals to a food source. They also use it to guide the swarm as a whole to a new home (Thomas D. Seeley and P. Kirk Visscher, "Choosing a Home: How the Scouts in a Honey Bee Swarm Perceive the Completion of Their Group Decision Making," *Behavioural Ecology and Sociobiology* 54 (2003): 511–520).

29 ***[Lindauer] looked closely at swarms flying overhead*** M. Lindauer, "Schwarmbienen auf Wohnungssuche," *Zeitschrift für vergleichende Physiologie* 37 (1955): 263–324.

30 ***other scientists confirmed his observation*** M. Beekman, R. L. Fathke, and T. D. Seeley, "How Does an Informed Minority of Scouts Guide a Honeybee Swarm as it Flies to Its New Home?" *Animal Behaviour* 71 (2006): 161–171).

30 ***a few informed individuals can lead uninformed individuals*** I. D. Couzin et al., "Effective Leadership and Decision Making in Animal Groups on the Move," *Nature* 455 (2005): 513–516.

30 ***Leadership "as a function of information differences"*** Madeleine Beekman, Gregory A. Sword, and Stephen J. Simpson, "Biological Foundations of Swarm Intelligence," in *Swarm Intelligence: Introduction and Applications*, edited by Christian Blum and Daniel Merkle (Berlin: Springer, 2008): 3–41. Many of the insect references in this chapter have been taken from this excellent review.

31 ***robots that will swarm around a human leader*** Hiroshi Hashimoto, "Cooperative Movement of Human and Swarm Robot Maintaining Stability of Swarm" *Proceedings of the 17th IEEE International Symposium on Robot and Human*

Interactive Communication, Technische Universität München, Munich, Germany, August 1–3 (2008): 249–254.

31 ***walk randomly in a circular room*** John R. G. Dyer et al., "Leadership, Consensus Decision Making and Collective Behavior in Humans," *Philosophical Transactions of the Royal Society B* 364 (2009): 781–789.

32 ***John C. Maxwell*** *The 21 Indispensable Qualities of a Leader: Becoming the Person Others Will Want to Follow* (Nashville: Thomas Nelson, 2007).

32 ***éminence grise*** The original *éminence grise* ("gray eminence") was the Capuchin friar François Leclerc du Tremblay, the right-hand man of Cardinal Richelieu. Capuchin friars wore brown robes, but for some obscure reason the color was referred to as "gray."

32 ***Dick Cheney, Edith Wilson, and Cardinal Wolsey*** See, respectively, Robert Kuttner, "Cheney's Unprecedented Power," *Boston Globe*, February 25, 2004, http://www.commondreams.org/views04/0225-01.htm; "The First Lady of the Land," *MitSprache* 20 (spring/summer 2008): http://www.cornelsen.de/sixcms/media.php/8/FirstLady_Mitsprache_S4-7_Eng.pdf?siteID=xyz; and "Cardinal Thomas Wolsey," *England Under the Tudors*, http://www.luminarium.org/encyclopedia/wolseybio.htm.

Ellis Parker Butler's wonderful character S. Potts explains to his fellow employees that he had never been elected president because he always prefers to be "the power behind the throne," the title of Butler's story (*Illustrated Sunday Magazine*, December 11, 1910, http://www.ellisparkerbutler.info/epb/biblio.asp?id=4303).

32 ***"the larger the group, the smaller the proportion of informed individuals needed"*** I. D. Couzin et al., "Effective Leadership and Decision Making in Animal Groups on the Move," *Nature* 455 (2005): 513–516.

32 ***Stanley Milgram arranged for groups*** S. Milgram, L. Bickman, and L. Berkowitz, "Note on the Drawing Power of Crowds of Different Size," *Journal of Personal and Social Psychology* 13 (1969): 79–82.

Milgram, a professor at Yale who died tragically young from heart failure in 1984 at the age of fifty-one, was famous for his experiments "to test how much pain an ordinary citizen would inflict on another person simply because he was ordered to by an experimental scientist." In his 1973 article "The Perils of Obedience" (*Harper's* [December]: http://scholar.google.co.uk/scholar?cluster=15613071320603007973&hl=en) he wrote that

> stark authority was pitted against the subjects' [participants'] strongest moral imperatives against hurting others, and, with the subjects' [participants'] ears ringing with the screams of the victims, authority won more often than not. The extreme willingness of adults to go to almost any lengths on the command of an authority constitutes the chief finding of the study and the fact most urgently demanding explanation.

33 ***Orpheus Chamber Orchestra provides a real-life example*** John Lubans Jr., "The Invisible Leader: Lessons for Leaders from the Orpheus Chamber Orchestra," *OD Practitioner* 38 (2006): 5–8, http://www.lubans.org/docs/odpissuesdownload-visitors.pdf.

33 ***six out of the thirty-one players sets the musical agenda*** The membership of this core is quite likely to vary from piece to piece!

34 ***"a leader is best when people barely know he exists"*** Although this proverb is often attributed to Laozi (Lao-tzu or Lao-tze), I have been unable to find it in his writings. The closest I could get was: "A good leader does not exercise authority. This is the value of unimportance." (*Tao Te Ching* chapter 68, http://www.chinapage.com/gnl.html#68).

Laozi himself is a semimythical figure, and according to the *Stanford Encyclopedia of Philosophy* may be a composite of several historical figures. The writings attributed to him may also be composites (http://plato.stanford.edu/entries/laozi/).

Professor Jens Krause has pointed out that leadership, whether invisible or not, involves a "cost" to the leader(s), since the demands of leadership can distract the leader and cause more accidents (Chantima Piyapong et al., "A Cost of Leadership in Humans," *Ethology* 113 [2007]: 821–824).

34 ***"I got off the plane in Rome at midnight"*** Personal communication, March 20, 2009.

Chapter 3

37 ***"go to the ant, thou sluggard"*** Proverbs 6:6, King James version.

37 ***"consider her ways"*** One of the "ways" of some ant species is to kidnap the larvae of smaller species and bring them up as slaves. Researchers have now found that the Spartacus-like response of the slaves is to mutiny, destroying two-thirds of the newly hatched queens and female workers of their captors (Alexandra Achenbach and Susanne Foitzik, "First Evidence for Slave Rebellion: Enslaved Ant Workers Systematically Kill the Brood of Their Social Parasite *Protomognathus americanus*," *Evolution* 63 [2009]: 1,068–1,075). The males of the kidnapping species do not participate in the slave raids, and male pupae were left alone. Food for thought!

37 ***Ants can distinguish objects up to three feet away*** Small wood ants (*Formica rufa*) can distinguish objects up to a foot-and-a-half away (S. P. D. Judd and T. S. Collett, "Multiple Stored Views and Landmark Guidance in Ants," *Nature* 392 [1998]: 710–714), and the larger (and more ferocious!) Australian bulldog ant can use its stereoscopic vision to distinguish objects up to three feet away (E. Sture Eriksson, "Attack Behaviour and Distance Perception in the Australian Bulldog Ant *Myrmecia Nigriceps*," *Journal of Experimental Biology* 119 [1985]: 115–131). Some ants may be able to use visual cues from even farther away (L. Passera, *L'organisation sociale des fourmis* [Toulouse: Privat, 1984]), but this still begs the question of how they do it when there are objects like rocks and twigs blocking their view.

37 ***Experiments on a laboratory colony of Argentine ants*** S. Goss, S. Aron, J. L. Deneubourg, and J. M. Pasteels, "Self-Organized Shortcuts in the Argentine Ant," *Naturwissenschaften* 76 (1989): 579–581.

38 ***the reason . . . is very obvious—once you think of it***

Sherlock Holmes said this in *The Adventure of the Copper Beeches.* He usually regretted explaining his thought processes. After one such discursion in *The Red Headed League*, his client Jabez Wilson "laughed heavily. 'Well, I never!' said he. 'I thought at first that you had done something clever, but I see that there was nothing in it after all.' 'I begin to think, Watson,' said Holmes, 'that I make a mistake in explaining. "Omne ignotum pro magnifico," you know, and my poor little reputation, such as it is, will suffer shipwreck if I am so candid.'"

39 ***"ant colony optimization"*** E. Bonabeau, M. Dorigo, and G. Theraulaz, "Inspiration for Optimization from Social Insect Behaviour," *Nature* 406 (2000): 39–42.

39 ***the optimal route that Ulysses might have taken*** Martin Grötschel and Manfred Padberg, "Ulysses 2000: In Search of Optimal Solutions to Hard Combinatorial Problems," ZIB-Report SC 93–34 (Berlin: Konrad-Zuse-Zentrum für Informationstechnik, 1993), www.zib.de/groetschel/pubnew/paper/groetschelpadberg1993.ps.gz. See also Karla Hoffman and Manfred Padberg, "Traveling Salesman Problem," http://iris.gmu.edu/~khoffman/papers/trav_salesman.html.

41 ***pedestrians [to establish paths] by habit*** Christopher Gray, "Streetscapes/Central Park's Bridle Paths; the Challenge of Restoring Long-Neglected Trails," *New York Times*, January 2, 1994.

41 ***Helbing and his colleagues have photographed and analyzed many such paths*** D. Helbing, J. Keltsch, and P. Molnár, "Modelling the Evolution of Human Trail Systems," *Nature* 388 (1997): 47–50. See also D. Helbing, "Pedestrian Dynamics and Trail Formation," in *Traffic and Granular Flow '97*, edited by M. Schreckenberg and D. E. Wolf (Singapore: National University of Singapore, 1998), 21–36. One of their many interesting observations is that the junction between two perpendicular paths tends to be in the form of a triangle (which shortens the distance), that is, Y-shaped rather than T-shaped.

41 ***Digg.com*** The analysis of the performance of the website was performed by Fang Wu and Bernardo A. Huberman from the Information Dynamics Laboratory at Hewlett-Packard

Laboratories, Palo Alto, California ("Novelty and Collective Attention," *Proceedings of the National Academy of Science of the USA* 104 [2007]: 17,599–17,601).

Wu and Huberman were mostly concerned with "the propagation of information in social networks, determining the effectiveness of advertising and viral marketing." Evolutionary biologist Simon Garnier noticed the relevance of their paper to ant foraging behavior and swarm intelligence and wrote about it on his blog ("Swarm Intelligence at digg.com," http://www.simongarnier.com/swarm-intelligence-at-diggcom/).

42 ***"ant colony routing"*** This process was first described by R. Schoonderwoerd et al. in "Ant-Based Load Balancing in Communications Networks," *Adaptive Behaviour* 5 (1997): 169–207.

43 ***ant colony routing can cope with these dynamic changes*** Gianni A. Di Caro, Frederick Ducatelle, and Luca M. Gambardella, "Theory and Practice of Ant Colony Optimization for Routing in Dynamic Telecommunications Networks," in *Reflecting Interfaces: The Complex Coevolution of Information Technology Ecosystems*, edited by N. Sala and F. Orsucci F. (Hershey, PA: Idea Group, 2008).

43 ***particle swarm optimization which emerged from the fertile minds*** J. Kennedy and R. C. Eberhart, "Particle Swarm Optimization," *Proceedings of the IEEE International Conference on Neural Networks* 4 (1995): 1,942–1,948.

See Ricardo Poli ("Analysis of the Publications on the Applications of Particle Swarm Optimisation," *Journal of Artificial Evolution and Applications* [2008], http://www.hindawi.com/journals/jaea/2008/685175.html.) for a very full list of examples and references. An excellent critical summary and other related problem-solving techniques, such as genetic algorithms (creating a random set of solutions and letting them fight it out, allowing "survival of the fittest" to determine the outcome) and Tabu search (searching closely related solutions for the best one, moving to this and then repeating the procedure—back-tracking not permitted) is given in Yanqiu Wang et al., "Survey of Modern

Optimization Techniques," *Advances in Information and Systems Science* 1 (2006): 108–118.

44 ***particle swarm optimization . . . for investment decision making*** J. Nenortaite and R. Simutis, "Adapting Particle Swarm Optimization to Stock Markets," in *Proceedings of the 5th International Conference on Intelligent Systems Design and Applications* (Washington, DC: IEEE Computer Society, 2005), 520–525.

44 ***analysis of MRI scans and satellite images*** M. Omran, A. Salman, and A. P. Engelbrecht, "Image Classification Using Particle Swarm Optimization" (lecture, 4th Asia-Pacific Conference on Simulated Evolution and Learning, Singapore, 2002).

45 ***tracking elephant migrations*** Parviz Palangour, Ganesh K. Venayagamoorthy, and Kevin Duffy, "Recurrent Neural Network Based Predictions of Elephant Migration in a South African Game Reserve" (lecture, International Joint Conference on Neural Networks, Vancouver, BC, July 16–21, 2006).

45 ***diagnosis of Parkinson's disease*** R. C. Eberhart and Xiaohui Hu, "Human Tremor Analysis Using Particle Swarm Optimization," *Evolutionary Computation* 3 (1999): 1,930.

45 ***plan as many right turns . . . as possible*** Brian Rooney, "UPS Figures Out the 'Right Way' to Save Money, Time and Gas: Delivery Giant Maximizes Efficiency with Ingenious Planning," *ABC News*, April 4, 2007, http://abcnews.go.com/wnt/Story?id=3005890&page=1.

46 ***Tests on volunteer swarms in the U.S. Navy*** Marc Kirschenbaum et al., "Human Swarm Testing of a Decision Support System for Cargo Movement Aboard Navy Ships," 2006, www.jcu.edu/math/swarm/papers/SIS2006.pdf.

46 ***"smart mob"*** Howard Rheingold, *Smart Mobs: The Next Social Revolution* (New York: Basic Books, 2002); Clive Thompson, "The Year in Ideas; Smart Mobs," *New York Times*, December 15, 2002, http://www.nytimes.com/2002/12/15/magazine/the-year-in-ideas-smart-mobs.html.

46 ***"smart mob rule"*** James McGirk, "Smart Mob Rule," *Foreign Policy* 92 (May–June 2003).

46 ***2001 demonstrations*** Clive Thompson, "The Year in Ideas; Smart Mobs," *New York Times*, December 15, 2002,

http://www.nytimes.com/2002/12/15/magazine/the-year-in-ideas-smart-mobs.html. Known as the EDSA demonstrations. They were named after Epifanio de los Santos Avenue, the road that rings the Philippine capital, Manila.

46 ***2005 civil unrest in France*** Patrick Hamon, the national police spokesman, said, "What we notice is that the bands of youths are, little by little, getting more organized" and are sending attack messages by mobile phone texts (Craig S. Smith, "As Rioting Spreads, France Maps Tactics," *New York Times* November 6, 2005, http://www.nytimes.com/2005/11/06/world/europe/06iht-france.html?pagewanted=all).

46 ***2006 student protests in Chile*** Jonathon Franklin, "Protests Paralyse Chile's Education System," *The Guardian*, June 7, 2006, http://www.guardian.co.uk/world/2006/jun/07/chile.schoolsworldwide.

46 ***Wild Strawberry student movement*** "All Students Around Taiwan Launch a 'Wild Strawberry Student Movement,'" *Taiwan News*, November 10, 2008, http://www.etaiwannews.com/etn/news_content.php?id=784881.

46 ***the Twitter network has been used to some effect to coordinate protests in Iran*** Lev Grossman, "Iran Protests: Twitter, the Medium of the Movement" *Time*, June 17, 2009, http://www.time.com/time/world/article/0,8599,1905125,00.html.

47 ***"more practical applications of Swarm Intelligence will continue to emerge"*** Simon Garnier, Jacques Gautrais, and Guy Theraulaz, "The Biological Principles of Swarm Intelligence," *Swarm Intelligence* 1 (2007): 29.

47 ***our brains themselves use the distributed logic of the ant colony*** Douglas Hofstadter, *Gödel, Escher, Bach: An Eternal Golden Braid* (Basic Books, New York, 1979); G. Buttazzo, "Artificial Consciousness: Utopia or Real Possibility?" *Computer* 34 (2001): 24–30.

In a rather beautiful twist, some MRI scanners use particle swarm optimization, one of the forms of distributed ant logic, to analyze the structure and activity of the brain itself! See, for example, M. P. Wachowiak et al., "An Approach to

Multimodal Biomedical Image Registration Utilizing Particle Swarm Optimization," *Evolutionary Computation* 8 (2004): 289–301.

Chapter 4

I am grateful to Dirk Helbing for providing me with an advance copy of his overview chapter with Anders Johansson, "Pedestrian, Crowd and Evacuation Dynamics," in *Encyclopedia of Complexity and Systems Science*, edited by Robert A. Meyers (New York: Springer, 2009). For details and images of the simulations discussed in this chapter, see Dirk Helbing, Illés Farkas, and Tamás Vicsek, "Simulating Dynamical Features of Escape Panic," *Nature* 407 (2000): 487–490, http://www.angel.elte.hu/~panic/.

49 ***Jerome K. Jerome discovered another way to make space*** *Three Men in a Boat* (London: Everyman's Library, 1957), 29. Out of deference to the reader, I have misquoted Jerome slightly. What he actually described was how passengers left the carriage hastily when a man "who seemed to belong to the undertaker class, said it [the smell] put him in mind of a dead baby."

50 ***three rules for boids*** With the addition of rules that specify the objectives of individuals within a crowd, these rules are still being used to model crowd behavior. See, for example, Benjamin Garrett et al., "Modeling Crowd Motion Using Swarm Heuristics and Predictive Agents," http://scholar.google.com.au/scholar?hl=en&lr=&cluster=4224582698595009105, and Martin Nygren, "Simulation of Human Behaviour in Stressful Crowd Situations," (master's thesis, computer science, Royal Institute of Technology, Stockholm, 2007 [abstract: http://www.nada.kth.se/utbildning/grukth/exjobb/rapportlistor/2007/sammanf07/nygren_martin.html]).

50 ***[Newton's] three laws of motion*** These appeared in Sir Isaac Newton's famous *Philosophiæ Naturalis Principia Mathematica* (usually known just as the *Principia*) in 1687. Some care is needed to state them in unambiguous, rigorous terms; the versions that I have quoted in the text

are simplified. A full statement would take us into the realms of Einsteinian physics and mass-energy conversion, but these would only be relevant if the crowds moved at somewhere near the speed of light.

There have been many demonstrations of the three laws on the Internet. One of the oddest was Las Vegas dancer Marie Celeste's tasteful striptease representation, which, unfortunately, is no longer available!

52 ***bundling physical and social forces*** This was first done in a pioneering paper by Dirk Helbing, Illés Farkas, and Tamás Vicsek: "Simulating Dynamical Features of Escape Panic," *Nature* 407 (2000): 487–490. A not-dissimilar approach was later used by Z. Fang, S. M. Lo, and J. A. Lu to calculate a relationship between crowd density and movement velocity in relation to the evacuation of a building ("On the Relationship Between Crowd Density and Movement Velocity," *Fire Safety Journal* 38 [2003]: 271–283). Curiously, the latter paper makes no mention of Helbing, Farkas, and Vicsek's paper, which had been published three years earlier.

52 ***[virtual crowd behavior] conformed to . . . video records of such crowds*** Dirk Helbing et al., "Self-Organized Pedestrian Crowd Dynamics: Experiments, Simulations, and Design Solutions," *Transportation Science* 39 (2005): 1–24; A. Johansson, D. Helbing, and P. K. Shukla, "Specification of the Social Force Pedestrian Model by Evolutionary Adjustments to Video Tracking Data," *Advances in Complex Systems* (October 2008): http://scholar.google.com.au/scholar?hl=en&q=P.+K.+Shukla+Specification+of+the+Social+Force+Pedestrian+Model+by+Evolutionary+Adjustments+to+Video+Tracking+Data+Advances+in+Complex+Systems+October+2008&btnG=Search.

Three other sources of data (unfortunately rather less accessible) are J. Kerridge and T. Chamberlain, "Collecting Pedestrian Trajectory Data in Real-Time," in *Pedestrian and Evacuation Dynamics '05*, edited by N. Waldau et al. (Berlin: Springer, 2005); S. P. Hoogendoorn, W. Daamen, and P. H. L. Bovey, "Extracting Microscopic Pedestrian Characteristics from Video Data," CD-ROM, in *Proceedings of the 82nd Meeting at the Transport Research Board* (Washington, DC: Mira Digital (2003); and K. Teknomo, "Microscopic Pedestrian Flow Characteristics: Development of an Image Processing Data Collection and Simulation Model" (Ph.D. dissertation, Tohoku University, Japan, 2002).

53 ***streams of pedestrians*** D. Helbing et al., "Self-Organizing Pedestrian Movement," *Environmental Planning B* 28 (2001): 361–383.

53 ***Army Ants organize themselves neatly into three-lane highways*** I. D. Couzin and N. R. Franks, "Self-Organized Lane Formation and Optimized Traffic flow in Army Ants," *Proceedings of the Royal Society B* 270 (2003): 139–146.

54 ***The net effect . . . is to slow the traffic down*** Scientists will recognize that the same mechanism underlies the phenomenon of viscosity in liquids, which basically arises from molecules hopping backward and forward between slower- and faster-moving streamlines in the liquid.

55 ***halve the rate of flow*** See figure 1(d) in Dirk Helbing, Illés Farkas, and Tamás Vicsek, "Simulating Dynamical Features of Escape Panic," *Nature* 407 (2000): 487–490, http://www.angel.elte.hu/~panic/.

55 ***"Once a pedestrian is able to pass through"*** See figure 3 in Dirk Helbing and Anders Johansson, "Pedestrian, Crowd and Evacuation Dynamics," in *Encyclopedia of Complexity and Systems Science*, edited by Robert A. Meyers (New York: Springer, 2009).

56 ***stripe formation*** K. Ando, H. Oto, and T. Aoki, "Forecasting the Flow of People," *Railway Research Review* 45 (1988): 8–13 (in Japanese); Dirk Helbing et al., "Self-Organized Pedestrian Crowd Dynamics: Experiments, Simulations, and Design Solutions," *Transportation Science* 39 (2005): 1–24.

57 ***"irregular succession of arch-like blockings"*** Dirk Helbing, Illés Farkas, and Tamás Vicsek, "Simulating Dynamical Features of Escape Panic," *Nature* 407 (2000): 487–490.

59 ***go with the crowd 60 percent of the time*** In Helbing et al.'s admittedly simplified picture, this strategy can reduce escape times from 43 seconds to 36 seconds in their model situation ("Self-Organized Pedestrian Crowd Dynamics:

Experiments, Simulations, and Design Solutions," *Transportation Science* 39 [2005]: 1–24).

59 ***our response is to often seek out family and friends*** Charles E. Fritz and Harry B. Williams, "The Human Being in Disasters: A Research Perspective," *Annals of the American Academy of Political and Social Science* 309 (1957): 42–51; Anthony R. Mawson, "Understanding Mass Panic and Other Collective Responses to Threat and Disaster," *Psychiatry* 68 (2005): 95–113.

This behavior applies even to companies of soldiers in battle conditions. Discipline, it turns out, is much less important than the presence of other soldiers who are known and trusted (Mawson, "Understanding Mass Panic," 101).

60 ***most of us are reluctant to accept and act upon warnings*** Charles E. Fritz and Harry B. Williams, "The Human Being in Disasters: A Research Perspective," *Annals of the American Academy of Political and Social Science* 309 (1957): 42.

60 ***Bill Bryson tells the story*** *The Life and Times of the Thunderbolt Kid* (London: Transworld Publishers, 2007), 258.

61 ***Marysville and Yuba, California . . . flood*** Charles E. Fritz and Harry B. Williams, "The Human Being in Disasters: A Research Perspective," *Annals of the American Academy of Political and Social Science* 309 (1957): 43.

61 ***"83 percent judged the situation to be very serious"*** Anthony R. Mawson, "Understanding Mass Panic and Other Collective Responses to Threat and Disaster," *Psychiatry* 68 (2005).

61 ***residents having gone back into the building*** J. Scanlon, "Human Behavior in a Fatal Apartment House Fire," *Fire Journal* (May 1979): 76–79, 122–123.

61 ***"panic, trample over each other, and lose all sense of concern"*** Charles E. Fritz and Harry B. Williams, "The Human Being in Disasters: A Research Perspective," *Annals of the American Academy of Political and Social Science* 309 (1957): 42.

61 ***One national columnist condemned them as barbarians*** Mike Royko, "The New Barbarians," *Cincinnati Post*, December 4, 1979.

61 ***"uncaring tread of the surging crowd"*** R. Burleigh, "Editors Notebook: At Death's Door," *Cincinnati Post*, December 8, 1979.

62 ***An analysis of statements taken by police*** Norris R. Johnson, "Panic at 'The Who Concert Stampede': An Empirical Assessment," *Social Problems* 34 (1987): 362–373.

63 ***crowd disaster that occurred*** Dirk Helbing, Anders Johannson, and Habib Zein Al-Abideen, "Dynamics of Crowd Disasters: An Empirical Study," *Physical Review E* (2007).

63 ***the multilevel Jamaraat Bridge*** When the crowd disaster occurred, it was just a two-level structure with a ground floor and a first floor, which is why it was called a bridge; it was only very recently replaced (2006–2007) by a multilevel building.

63 ***Modern image analysis techniques*** The software, called Crowd Vision, was designed by Anders Johansson, from the Eldenössische Technische Hochschule (ETH) University in Zurich, and it won him the inaugural Crowded Places Idea competition award in 2008. The software automatically detects and tracks pedestrians while they move under CCTV cameras. "The concept," according to Johansson, "is to run the analysis in real time of multiple CCTV cameras to get a more accurate view of how crowds are moving. This allows for a certain amount of forecasting of how a situation will develop—thereby offering the ability to use a range of interventions before a situation becomes critical" ("Crowd-Vision Technology Wins $10,000 'Crowded Places' Award at GSC, Global Security Challenge," November 12, 2008, http://www.globalsecuritychallenge.com/blog_detail.php?id=220).

64 ***The mathematics that force chains describes*** One similarity is that the displacements tend to follow a power law (see P. A. Johnson and X. Jia, "Nonlinear Dynamics, Granular Media and Dynamic Earthquake Triggering," *Nature* 437 [2005]: 871–874).

64 ***credit of the Saudi Arabian authorities*** Dirk Helbing, Anders Johannson, and Habib Zein El-Abideen, "Dynamics of Crowd Disasters: An Empirical Study," *Physical Review E*

(2007). See also the supplement to the paper at http://www.trafficforum.org/crowdturbulence.

Chapter 5

This chapter benefited greatly from the ideas, writings, and comments of Michael Mauboussin, chief investment strategist at Legg Mason Capital Management in Baltimore and adjunct professor at Columbia Business School in New York. Thanks very much, Michael!

67 ***a plethora of examples*** James Surowiecki, *The Wisdom of Crowds: Why the Many Are Smarter Than the Few and How Collective Wisdom Shapes Business, Economies, Societies, and Nations* (New York: Anchor, 2005).

68 ***a wombat or a wallaby*** Wombats are compact hairy marsupials that were described by my father as "hairy steamrollers." They weigh between 40 and 80 pounds. The wallabies in the region where we were are called "swamp wallabies." They stand around 3 feet high and weigh up to 37 pounds.

69 ***many wrongs can come pretty close*** When birds such as skylarks, ducks, and homing pigeons fly together in flocks, they take advantage of this principle. The flock as a whole flies in a direction that is the average of the directions chosen by its individual members. Each of those directions has some degree of error, but the bigger the flock, the more the errors cancel out, and the more accurate the navigation (Andrew M. Simons, "Many Wrongs: The Advantages of Group Navigation," *Trends in Ecology and Evolution* 19 [2004]: 453–455).

70 ***the power of prayer*** Francis Galton, "Statistical Inquiries into the Efficacy of Prayer," *Fortnightly Review New Series* 68 (1872): 125–135.

70 ***"in these democratic days"*** Francis Galton, "Vox Populi," *Nature* 75 (1907): 450–451.

70 ***Fat Stock and Poultry Exhibition . . . guessing competition*** Galton realized that the data were fit for statistical analysis because "the judgments were unbiased by passion and uninfluenced by oratory and the like. The sixpenny fee

deterred practical joking, and the hope of a prize and the joy of competition prompted each competitor to do his best" (Francis Galton, "Vox Populi," *Nature* 75 [1907]: 450).

71 ***"one vote, one value"*** Francis Galton, "One Vote, One Value," *Nature* 75 (1907): 414.

71 ***complexity scientist Scott Page*** Scott Page, *The Difference: How the Power of Diversity Creates Better Groups, Firms, Schools, and Societies* (Princeton: Princeton University Press, 2007).

71 ***Karl Pearson recalculated Galton's result*** Karl Pearson, ed., *The Life, Letters and Labours of Francis Galton: Researches of Middle Life* (Cambridge: Cambridge University Press, 1924), 404–405, http://galton.org/cgi-bin/search Images/galton/search/pearson/vol2/pages/vol2_0468.htm.

71 ***[Galton] believed that the median was the right one to use*** David M. Levy and Sandra J. Peart argue that Surowiecki's quoting of this figure as the one that Galton believed in is misleading, to say the least ("The Tale of Galton's Mean: The Influence of Experts," October 4, 2008, http://adamsmithlives.blogs.com/thoughts/files/tale_of_g altons_mean.pdf). I agree; Galton always thought that the median was the correct one to use. He did report the mean in a reply to a reader's letter (*Nature* 75 [1907]: 509–510), but he took pains to say that the median was the better estimate. He was well aware that a mean has meaning only if the results are uniformly distributed above and below it (as in the famous bell curve, which represents the distribution of many characteristics; see for example Richard J. Herrnstein and Charles Murray *The Bell Curve: Intelligence and Class Structure in American Life* [New York: Free Press, 1996]).

According to Levy and Peart, "by plotting the actual guesses against a hypothetical normal distribution, Galton found striking evidence of non-normality. This supported his statistical intuition that the mean would be a poor measure of central tendency or 'interpretation of their collective view'"—and so he chose the median as being more meaningful.

71 ***London architect Matt Deacon . . . took a glass jar*** Matt Deacon, "The Ox, the Jam Jar and the Architect," May

18, 2007, http://blogs.technet.com/matt_deacon/archive/2007/05/18/the-ox-the-jam-jar-and-the-architect.aspx.

The mean value was 419, but the median (calculated from Deacon's original data, which he kindly provided in a personal communication) was 335.

72 ***Wall Street investment strategist Michael Mauboussin . . . tested the ability of students . . . to etimate*** Michael Mauboussin, "Explaining the Wisdom of Crowds: Applying the Logic of Diversity," Legg Mason Capital Management, March 20, 2007, http://www.adamdell.com/Documents/ExplainingWisdom.pdf.

72 ***Joe Nocera compared [it] . . . to the situation in* Shakespeare in Love** "The Future Divined by the Crowd," *New York Times*, March 11, 2006.

73 ***Page's diversity prediction theorem*** Scott Page, *The Difference: How the Power of Diversity Creates Better Groups, Firms, Schools, and Societies* (Princeton: Princeton University Press, 2007), 208. Here's an example of how it works:

Let's say there are three of us standing on a street corner and trying to guess how many green cars will come down the street in the next five minutes. One of us guesses 5, one guesses 10, and one guesses 15. As it turns out, 12 green cars come through in that time.

Our collective error is the difference between our average answer and the real answer. Our average answer is 10. The real answer is 12, so our collective error = 2, and our squared collective error is 4.

Our average individual error is the average of all of our squared errors. The first person was 7 cars out, the second was out by 2, and the third was out by 3. So our average squared individual error is

$$(7^2 + 2^2 + 3^2)/3 = 62/3 = 20.66$$

Our prediction diversity is the scatter of individual guesses, worked out as the average difference between our individual guesses and the average of our individual guesses. It sounds complicated, but it's dead simple. The average of our individual guesses was $(5 + 10 + 15)/3 = 10$. The first person was 5 away from that, the second was spot on, and the third was 5 away. So our squared prediction diversity is $(5^2 + 0^2 + 5^2)/3 = 50/3 = 16.66$

Let's check it out. According to Page's theorem:

(squared) collective error = average (squared) individual error–(squared) prediction diversity

$$4 = 20.66 - 16.66$$

Bingo!

Some people have argued that the many wrongs principle simply arises from a well-known result in statistics, in which the standard error in a mean decreases with the square root of the number of observations. In his classic paper on the jelly bean experiment and market efficiency, for example, Jack Treynor suggests the model's accuracy "comes from the faulty opinions of a large number of investors who err independently. If their errors are wholly independent, the standard error in equilibrium price declines with roughly the square root of the number of investors" ("Market Efficiency and the Bean Jar Experiment," *Financial Analysts Journal* [May–June 1987]).

Michael Mauboussin argues that this interpretation is faulty.

> We believe that the square-root law, which says the standard error of the mean decreases with the square root of N (number of observations), is an inappropriate explanation for the jelly bean (or market efficiency) problem. The square-root law applies to sampling theory, where there are independent observations that include the answer plus a random noise term. Over a large number of observations, the errors cancel out. An example is observing and measuring star luminosity. The underlying assumption behind the square-root law is the observations are independent and identically distributed around a mean. This is clearly not the case with either the jelly bean jar or markets. We believe the diversity prediction theorem is a more robust way

to explain the wisdom of crowds in this case. (Michael Mauboussin, "Explaining the Wisdom of Crowds: Applying the Logic of Diversity," Legg Mason Capital Management, March 20, 2007, http://www.adamdell.com/Documents/ExplainingWisdom.pdf).

74 ***cognitive diversity*** Michael Mauboussin, "Explaining the Wisdom of Crowds: Applying the Logic of Diversity," Legg Mason Capital Management, March 20, 2007, 2, http://www.adamdell.com/Documents/ExplainingWisdom.pdf.

74 ***"being different is as important as being good"*** Scott Page, *The Difference: How the Power of Diversity Creates Better Groups, Firms, Schools, and Societies* (Princeton: Princeton University Press, 2007), 208.

75 ***the advantage of diversity*** In our local quiz team, for example, we always try to have one person who knows a lot about film and television, another with a broad knowledge of history, and someone who understands geography, because these three categories often come up.

74 ***a diverse set of employees . . . can forecast product sales and profits*** Quoted from Michael Mauboussin, "What Good Are Experts?" *Harvard Business Review* (February 2008). See also Michael Mauboussin, "The Importance of Diverse Thinking: Why the Santa Fe Institute Can Make You a Better Investor," January 16, 2007, http://www.leggmasoncapmgmt.com/pdf/diversethinking1.pdf

74 ***Page gives the example of a group of football journalists*** Scott Page, *The Difference: How the Power of Diversity Creates Better Groups, Firms, Schools, and Societies* (Princeton: Princeton University Press, 2007), 210–212.

75 ***experts come into their own*** Michael Mauboussin, "The Importance of Diverse Thinking: Why the Santa Fe Institute Can Make You a Better Investor," January 16, 2007, http://www.google.co.uk/search?hl=en&client=firefox-a&channel=s&rls=org.mozilla%3Aen-US%3Aofficial&hs=Zqk&q=Michael+Mauboussin+The+Importance+of+Diverse+Thinking%3A+Why+the+Santa+Fe+Institute+Can+Make+You+a+Better+Investor%2C%E2%80%9D&btnG=Search&meta=.

76 ***The group as a whole . . . got 11 out of 12 right*** The one that got the most votes (i.e., the modal, or majority, selection) within each category was counted as the group selection (Michael Mauboussin, "Explaining the Wisdom of Crowds: Applying the Logic of Diversity," Legg Mason Capital Management, March 20, 2007, 6, http://www.adamdell.com/Documents/ExplainingWisdom.pdf).

76 ***Franklin and Jefferson both spent time in Paris*** As commissioner for the United States, Franklin visited Paris several times between 1776 and 1785, and as U.S. minister to France, Jefferson visited several times between 1784 and 1789.

76 ***the first French constitution*** This document has been replaced, and there have been several subsequent versions, the last being adopted on October 4, 1958. This Constitution of the Fifth Republic has already been amended eighteen times, and there are further amendments in the pipeline at the time of this writing.

It is interesting to compare the French Declaration of the Rights of Man with the U.S. Declaration of Independence.

The French Declaration (August 26, 1789) begins:

> 1. Men are born and remain free and equal in rights. Social distinctions may be founded only upon the general good. 2. The aim of all political association is the preservation of the natural and imprescriptible rights of man. These rights are liberty, property, security, and resistance to oppression. 3. The principle of all sovereignty resides essentially in the nation. No body nor individual may exercise any authority which does not proceed directly from the nation. 4. Liberty consists in the freedom to do everything which injures no one else; hence the exercise of the natural rights of each man has no limits except those which assure to the other members of the society the enjoyment of the same rights. These limits can only be determined by law. (Yale Law School Avalon Project, http://avalon.law.yale.edu/18th_century/rightsof.asp)

The second sentence of the U.S. Declaration of Independence (July 4, 1776) reads more simply: "We hold these truths to be self-evident, that all men are created equal, that they are endowed by their Creator with certain unalienable Rights, that among these are Life, Liberty and the pursuit of Happiness." (http://www.archives.gov/exhibits/charters/declaration_transcript.html)

76 ***the Marquis de Condorcet*** Pronounced "con-dor-SAY."

76 ***Franklin met up with Condorcet*** Franklin and Jefferson must also have met Condorcet's wife, Sophie, whose memorable maiden name was de Grouchy. A historical account of their meeting, and of the possible effects of Condorcet's jury theorem on the framing of the American Constitution, is given by Nicholas Quinn Rosenkranz in "Condorcet and the Constitution: A Response to *The Law of Other States*," *Stanford Law Review* 59 (2007): 1,281–1,308.

Condorcet's ideas were initially disseminated in America by Nicholas Collin, one of the eighteenth-century pioneers of the American Philosophical Society (Arnold B. Urken and Iain McLean, "Nicholas Collin and the Dissemination of Condorcet in the United States," *Science in Context* 20 [2007]: 125–132).

77 ***[Condorcet's social mathematics] "had to be discussed"*** Nicholas Quinn Rosenkranz, "Condorcet and the Constitution: A Response to *The Law of Other States*," *Stanford Law Review* 59 (2007): 1,281–1,308.

77 ***Condorcet's remarkable work*** *Essai sur l'application de l'analyse a la probabilitie des decisions rendues a la pluralite des voix* (Paris: L'Impremerie Royale, 1785), translated by Iain McLean and Fiona Hewitt as *Condorcet: Foundations of Social Choice and Political Theory* (London: Edward Elgar Publishing, 1994).

77 ***"Condorcet's jury theorem"*** See, for example, P. J. Borland, "Majority Systems and the Condorcet Jury Theorem," *Statistician* 38 (1989): 181–189.

77 ***John Adams told Jefferson that Condorcet was a "mathematical charlatan"*** Arnold B. Urken and Iain

McLean, "Nicholas Collin and the Dissemination of Condorcet in the United States," *Science in Context* 20 (2007): 128.

77 ***the chance . . . rapidly becomes closer to 100 percent*** The unfortunate corollary to this conclusion is that if the individual members of a group have a *less* than 50:50 chance of getting the right answer, then the chance of the majority verdict being right *decreases* dramatically when the size of the group increases. The best chance of finding the right answer under these conditions is to accept the opinion of a single group member and hope for the best!

Jefferson seems to have ignored Adams' opinion and got on with Condorcet particularly well because both of them believed that the scientific method could be applied to politics. Condorcet coined the term *science politique*, which Jefferson seems to have been the first to translate as "political science." Jefferson and Franklin both realized, though, that the science that Condorcet was peddling was more mathematical and philosophical than political and realistic. So did the framers of the first French constitution, which happened at very nearly the same time.

The dates of the two constitutions actually overlapped. The Constitution of the United States of America was completed on September 17, 1787, and ratified by the thirteen states thereafter, with Rhode Island being the last on May 29, 1790 (by the desperately close vote of 34 to 32). The first French constitution was adopted on October 6, 1789, following the Declaration of the Rights of Man of August 26 of that year.

78 ***"increasing the number of legislative bodies"*** The Marquis de Condorcet quoted in Nicholas Quinn Rosenkranz in "Condorcet and the Constitution: A Response to *The Law of Other States*," *Stanford Law Review* 59 (2007): 1,293.

78 ***the U.S. Supreme Court made . . . a 1983 judgment about the functions of the two houses*** *INS vs. Chadha* 462 U.S. 919, 950 (1983), quoted in Nicholas Quinn Rosenkranz in "Condorcet and the Constitution: A Response

to *The Law of Other States,*" *Stanford Law Review* 59 (2007): 1,295.

79 ***Condorcet suggested that Louis XVI's jury be set up in this way*** Arnold B. Urken and Iain McLean, "Nicholas Collin and the Dissemination of Condorcet in the United States," *Science in Context* 20 (2007): 126.

80 ***"folks with nothing better to do on a weekday afternoon"*** Michael Mauboussin, "Explaining the Wisdom of Crowds: Applying the Logic of Diversity," Legg Mason Capital Management, March 20, 2007, 3, http://www.adamdell.com/Documents/ExplainingWisdom.pdf.

80 ***even if only a few people know the answer and the rest are guessing*** We saw this in chapter 3 when just a few knowledgeable people in a crowd were able to lead the rest to a predetermined destination.

80 ***try the following question . . . on your friends*** This example was suggested by Scott Page in *The Difference: How the Power of Diversity Creates Better Groups, Firms, Schools, and Societies* (Princeton: Princeton University Press, 2007).

80 ***28 votes for Noll and 24 for each of the others*** Statistical scatter makes this difference less meaningful, but with increasing group size such differences between means become more significant, because the distribution approaches more closely to a smooth, "normal" one. This is known as the "central limit theorem." For an example that involves rolling virtual dice, see http://www.stat.sc.edu/~west/javahtml/CLT.html. For a brave and largely successful effort to explain the theorem without getting into too much mathematical detail, see http://www.intuitor.com/statistics/CentralLim.html.

Chapter 6

83 ***Consensus: A Foolish Consistency?*** The title of this chapter comes from Ralph Waldo Emerson's well-known saying: "A foolish consistency is the hobgoblin of little

minds" ("Self-Reliance," *Essays*, Merrill's English Texts [New York: Charles E. Merrill Co., 1907]). The phrase has gained currency in management circles through the technique of consistency management (Anthony Finkelstein, "A Foolish Consistency: Technical Challenges in Consistency Management," *Lecture Notes in Computer Science* [London: Springer, 2000]), which applies particularly to the collaborative development of computer software.

83 ***I was involved in the formation of a new political party*** It was called the Australia Party (formed in 1969).

83 ***Delphi technique*** For a simple description, see http://creatingminds.org/tools/delphi.htm. The technique is described more fully in my book *Rock, Paper, Scissors: Game Theory in Everyday Life* (New York: Basic Books, 2008), 51–52.

84 ***Samuel Johnson . . . "took care that the Whig dogs did not get the best of it"*** This well-known story is based on a comment that Johnson made at a dinner party (J. P. Hardy, *Samuel Johnson: A Critical Study* [New York: Routledge, 1979], 42), but in fact Johnson was very evenhanded in his approach (Steven D. Scherwatzky, "Complicated Virtue: The Politics of Samuel Johnson's Life of Savage," *Eighteenth-Century Life* 25 [2001]: 80–93), despite his personal preference for the more conservative policies of the aristocratic Tories. There was also no tradition of reporting speeches verbatim, and Johnson's cavalier approach was regarded more with admiration than criticism. When Philip Francis, the translator of Demosthenes, remarked at the same party that one of Pitt's speeches was the best he had ever read, he was even more impressed after Johnson admitted having written it "in a garret in Exeter Street" (Hardy, *Samuel Johnson*, 42).

85 ***a clue that the people of Moscow used*** The people of Great Britain also used it during the austere times after World War II. The poet Philip Larkin, then studying at Oxford, reports that "it became a routine after ordering one's books in the Bodley [library] after breakfast to go and look for a cake or cigarette queue" (*Required Writing* [London: Faber & Faber, 1983]).

85 ***quorum response*** This refers to "an animal's probability of exhibiting a behaviour is a sharply non-linear function of the number of other animals already exhibiting the behaviour." It is based on the definition of a human quorum as "a gathering of the minimal number of members of an organization to conduct business." (http://wordnetweb.princeton.edu/perl/webwn). The phrase in connection with animal behavior was coined by David J. T. Sumpter and Stephen C. Pratt in "Quorum Responses and Consensus Decision Making," *Philosophical Transactions of the Royal Society B* 364 (2009): 743–753. The paper is one of a series of eleven for a special issue on "group decision making in humans and animals."

85 ***the neurons in the human brain show a similar sort of response*** Many questions remain to be answered about such responses. See Iain D. Couzin, "Collective Cognition in Animal Groups," *Trends in Cognitive Sciences* 13 (2008): 36–43.

86 ***make a choice between speed and accuracy*** S. C. Pratt and D. J. T. Sumpter, "A Tunable Algorithm for Collective Decision-Making," *Proceedings of the National Academy of Sciences of the U.S.A.* 103 (2006): 15,906–15,910.

87 ***lemmings filmed while cascading over a cliff in the film* White Wilderness** This film helped to promote the myth that the lemmings are committing suicide. That interpretation of the scene is, of course, false. When a large migrating group reaches a cliff edge, pressure from those behind can push the front ones over the edge. Also, lemmings *can* swim, and they drown only if they become exhausted.

87 ***interdependence*** This term means "individuals making choices that rely on the decisions of others" (David J. T. Sumpter and Stephen C. Pratt, "Quorum Responses and Consensus Decision Making," *Philosophical Transactions of the Royal Society B* 364 [2009]: 743). Interdependence is the basis of group decision making in nature. Experiments suggest that we have a built-in tendency to base our decisions about how to act on the decisions and actions of others. In Stanley Milgram's staring up at nothing experiment

(p. 32), for example, just one person staring up at an empty sixth-floor window could induce some 40 percent of passersby to stare with him. When two people were staring up, the proportion of copiers rose to 60 percent. With five it went up to 80 percent.

This was not, however, a quorum response. In a quorum response, the chance of a passerby looking up would have more than doubled when there were two people staring up instead of one. The response was more like the behavior of capuchin monkeys who are traveling in a troop: the probability of a monkey copying its neighbor's choice of direction simply doubles if twice as many neighbors have chosen that direction (H. Meunier et al., "Group Movement Decisions in Capuchin Monkeys: The Utility of an Experimental Study and a Mathematical Model to Explore the Relationship Between Individual and Collective Behaviors," *Behaviour* 143 [2007]: 1,511–1,527).

Instead of using a quorum response, the monkeys are guided by a mixture of dominance hierarchies, previous experience, and social interactions (S. Garber and P. A. Boinski, *On the Move: How and Why Animals Travel in Groups* [Chicago: University of Chicago Press, 2000]). Rather like us, really.

87 ***informational cascade*** S. Bikhchandani, D. Hirshleifer, and I. Welch, "A Theory of Fads, Fashions, Custom and Cultural Change as Informational Cascades," *Journal of Political Economy* 100 (1992): 992–1,026.

87 ***cockroaches, ants, and spiders . . . use the quorum response*** David J. T. Sumpter and Stephen C. Pratt, "Quorum Responses and Consensus Decision Making," *Philosophical Transactions of the Royal Society B* 364 (2009): 743–753. The chance of a cockroach leaving a site of shelter to search for a better one, for example, decreases sharply with the number of other cockroaches using the original site (J. M. Ame et al., "Collegial Decision Making Based on Social Amplification Leads to Optimal Group Formation," *Proceedings of the National Academy of Sciences of the U.S.A.* 103 [2006]: 5,835–5,840).

88 ***able to detect a lie instantly*** Paul Ekman, Maureen O'Sullivan, and Mark G. Frank, "A Few Can Catch a Liar,"

Psychological Science 10 (1999): 263–266. They seem to do it by picking up microexpressions, which were first identified by Ekman from video recordings (Paul Ekman, *Telling Lies: Clues to Deceit in the Marketplace, Politics, And Marriage*, 2nd ed. [New York: Norton, 1992]). Ekman has pursued this theme in subsequent work (e.g., *Emotions Revealed: Recognizing Faces and Feelings to Improve Communication and Emotional Life*, 2nd ed. [New York: Owl Books, 2007]). His work is the basis for the popular TV show "Lie to Me" (http://www.imdb.com/title/tt1235099/; www.paulekman.com).

88 ***the 0.25 percent of people that scientists have found can . . . [detect lies] almost* all *the time*** Gregory A. Perez, "'Wizards' Can Spot the Signs of a Liar: A Rare Few Have the Skill to Detect Flickers of Falsehood, Scientists Say," Associated Press, October 14, 2004, http://www.msnbc.msn.com/id/6249749/.

87 ***the interplay of independence and interdependence that gives us our best chance*** Bees also use it very successfully when searching for a new nest site. A search committee of several hundred scouts inspect potential nest sites and then return to perform the well-known waggle dance. Initially, their search is random and their dances are independent. Once dancing activity has built up, though, they are more likely to inspect sites advertised by others. Eventually, high-quality sites become more and more visited, and there are longer dances for these sites until finally there is a consensus (Christian List, Christian Elsholtz, and Thomas D. Seeley, "Independence and Interdependence in Collective Decision Making: An Agent Based Model of Nest-Site Choice by Honeybee Swarms," *Philosophical Transactions of the Royal Society B* 364 [2009]: 755–762).

93 ***social pressures within the group push its members*** Another way of putting it is that groupthink arises when the pressures of group members on one another narrow down the range of opinions (David J. T. Sumpter and Stephen C. Pratt, "Quorum Responses and Consensus Decision Making," *Philosophical Transactions of the Royal Society B* 364 [2009]: 743–753).

89 ***an annual process of negative voting*** S. Forsdyke, *Exile, Ostracism and Democracy: The Politics of Expulsion in Ancient Greece* (Princeton: Princeton University Press, 2005). See also Paul Cartledge, "Ostracism: Selection and De-Selection in Ancient Greece," http://www.historyand policy.org/papers/policy-paper-43.html.

89 ***voting paradox*** See, for example, Hanna Nurmi, *Voting Paradoxes and How to Deal with Them* (New York: Springer, 1999).

Condorcet's voting paradox is not to be confused with the paradox of voting, although the two terms are sometimes used interchangeably (e.g., Robert M. May, "Some Mathematical Remarks on the Paradox of Voting," *Behavioral Science* 16 [1970]: 143–151; an article that is actually about the rock-paper-scissors voting paradox conundrum). The paradox of voting, first put forward by Anthony Downs in *An Economic Theory of Democracy* (New York: Harper Collins, 1957), refers to the problem of why people bother to vote at all. The argument is that an individual's vote is very unlikely to swing an election, and since it takes an effort to go out and vote, it is hardly worth the effort. Yet many people go out and vote—that's the paradox.

90 ***presented them to a class of fourth graders*** Donald G. Saari, "A Fourth-Grade Experience," August 1991, http://www.colorado.edu/education/DMP/voting_c.html.

91 ***An unwelcome further complexity was discovered by . . . Kenneth Arrow*** K. J. Arrow, "A Difficulty in the Concept of Social Welfare," *Journal of Political Economy* 58 (1950): 328–346; Kenneth J. Arrow, *Social Choice and Individual Values*, 2nd ed. (New York: Wiley, 1963).

92 ***"implications of the paradox of social choice"*** http://nobelprize.org/nobel_prizes/economics/laureates/1972/arrow-lecture.html.

92 ***"democracy is the worst form of government"*** Sir Winston Churchill, *Hansard* (transcripts of the U.K. Parliamentary debates) November 11, 1947, http://hansard.mill banksystems.com/.

93 ***First past the post*** When I was typing this, it came out as "fist past the post," which seems an appropriate mistype in

view of the fact that it is the power of the major parties that keeps this anachronistic system in place.

93 ***People may vote tactically*** And they may do it even when they are all on the same side. For an up-to-date discussion of the "surprisingly subtle interaction between the voting rules used to make decisions and the incentives for committee members to share information prior to voting," see David Austen-Smith and Timothy J. Feddersen, "Information Aggregation and Communication in Committees," *Philosophical Transactions of the Royal Society B* 364 (2009): 763–769.

93 ***"a pattern of thought that is characterized by self-deception"*** *Merriam-Webster's Collegiate Dictionary*, 11th ed. (Springfield, MA: Merriam-Webster, 2003), http://www.merriam-webster.com/dictionary/groupthink.

94 ***MAD (mutually assured delusion)*** This wonderful acronym was coined by the Princeton economist and public affairs expert Roland Bénabou ("Groupthink and Ideology," [Schumpeter lecture, 22nd Annual Congress of the European Economics Association, Budapest, August 2007]).

94 ***fall "prey to a collective form of overconfidence and willful blindness"*** Roland Bénabou, "Groupthink and Ideology" (Schumpeter lecture, 22nd Annual Congress of the European Economics Association, Budapest, August 2007).

94 ***Irving Janis coined the term* groupthink *in 1972*** Irving L. Janis, *Victims of Groupthink: A Psychological Study of Foreign-Policy Decisions and Fiascoes* (Boston: Houghton Mifflin, 1972).

Janis' arguments were largely psychological. Princeton economist Roland Bénabou expanded them and put them on a firmer footing with an argument that has roots in game theory:

> Whenever an agent benefits (on average) from others' delusions, this tends to make him more of a realist; and whenever their disconnection from reality makes him worse off this pushes him towards denial, which is contagious. This "psychological multiplier" [effect] . . . implies that, in organizations where some agents (e.g.,

managers) have a greater impact on others' welfare (e.g., workers) than the reverse, strategies of realism or denial will "trickle down" the hierarchy, so that subordinates will in effect take *their beliefs from their leader(s)* [because they benefit more than they would if they did the reverse]. (Roland Bénabou, "Groupthink and Ideology" [Schumpeter lecture, 22nd Annual Congress of the European Economics Association, Budapest, August 2007]).

In other words, strategies in response to behaviors at the top work their way through an organization through a sort of chain reaction.

94 ***According to investigative reporter Bob Woodward*** Quoted in Derek Kravitz, "Bush's Legacy in Iraq," Washington Post Investigations, *Washington Post*, January 13, 2009, http://voices.washingtonpost.com/washingtonpostinvestigations/2009/01/iraq_war_defining_legacy_of_bu.html.

99 ***"very high echelon groups . . . intoxicating levels"*** Irving L. Janis, quoted in Robert S. Baron, "So Right It's Wrong: Groupthink and the Ubiquitous Nature of Polarized Group Decision-Making," in *Advances in Experimental Social Psychology*, edited by Leonard Berkowitz and Mark P. Zanna, vol. 37 (Maryland: Elsevier Academic Press, 2005), 222.

99 ***Groupthink is everywhere*** Robert S. Baron, "So Right It's Wrong: Groupthink and the Ubiquitous Nature of Polarized Group Decision-Making," in *Advances in Experimental Social Psychology*, edited by Leonard Berkowitz and Mark P. Zanna, vol. 37 (Maryland: Elsevier Academic Press, 2005), 219–253.

This even applies to the rarefied and supposedly objective world of academia. A colleague of mine, a leading researcher in statistics, could not work out why he was consistently being turned down for senior professorial positions. His research was highly original, but it was unorthodox and outside the mainstream. All of the individual selection committee members he spoke to told him that his research was excellent. When they got together in committee, however, no one was willing to speak up in support of it.

94 ***main characteristics [of groupthink]*** I. L. Janis and L. Mann, *Decision-Making: A Psychological Analysis of Conflict, Choice and Commitment* (New York: Free Press, 1977).

95 ***Feynman's investigations*** These are reported in his book *What Do YOU Care What Other People Think? Further Adventures of a Curious Character* (New York: Norton, 1988).

95 ***"there are enormous differences of opinion"*** See, for example, Feynman's appendix to the report of the Rogers Commission on the *Challenger* disaster, "Appendix F: Personal Observations on the Reliability of the Shuttle," http://science.ksc.nasa.gov/shuttle/missions/51-l/docs/rogers-commission/Appendix-F.txt.

96 ***[Feynman] dunked it in the glass of ice water*** Richard P. Feynman, *What Do YOU Care What Other People Think? Further Adventures of a Curious Character* (New York: Norton, 1988), 151–152. A video was apparently made of Feynman's demonstration, but there is unfortunately no copy on YouTube.com. See, however, *Challenger: The Untold Story*, http://www.youtube.com/watch?v=xV25ol-NedQ.

96 ***[Feynman was] permitted to add an appendix*** Richard P. Feynman, "Appendix F: Personal Observations on the Reliability of the Shuttle," http://science.ksc.nasa.gov/shuttle/missions/51-l/docs/rogers-commission/Appendix-F.txt.

97 ***the underlying problem which led to the Challenger accident*** See http://www.gpoaccess.gov/challenger/index.html under "Space Shuttle Accident."

97 ***space shuttle* Columbia *disaster*** NASA has released a video on YouTube.com: "Space Shuttle *Columbia* Disaster from NASA TV," http://www.youtube.com/watch?v=LijS7XP4vp8.

98 ***a panel of scientists . . . to "brainstorm"*** Groupthink in brainstorming groups can also influence the way in which the members think of themselves compared to the members of other similar groups. In one study, students were asked to brainstorm a problem in groups and then asked to evaluate the performance of their group against those of others. With-

out exception, each group rated its own performance the best. (K. M. Long and R. Spears, "Opposing Effects of Personal and Collective Self-Esteem on Interpersonal and Intergroup Comparisons," *European Journal of Social Psychology* 28 [1998]: 913–930. This rather difficult article is summarized succinctly in Robert S. Baron, "So Right It's Wrong: Groupthink and the Ubiquitous Nature of Polarized Group Decision-Making," in *Advances in Experimental Social Psychology*, edited by Leonard Berkowitz and Mark P. Zanna, vol. 37 [Baltimore: Elsevier Academic Press, 2005], 219–253).

98 ***I was never invited back*** Following the example of Feynman, I have now turned to writing my own independent "Appendix F" in the form of books, articles, and broadcasts that are designed to let people know how science *really* works, rather than how administrators think it ought to work.

99 ***They asked volunteer Hindu and Muslim students . . . to read stories*** Donald Taylor and Vaishna Jaggi, "Ethnocentrism and Causal Attribution in the South Indian Context," *Journal of Cross-Cultural Psychology* 5 (1974): 162–171.

100 ***[Australian] teenager Corey Delaney*** See, for example, "Aussie Party Boy Corey Delaney Finds Stardom Online," *The Times*, January 17, 2008, http://www.timesonline.co.uk/tol/news/world/article3198418.ece.

100 ***a publicity agent sought Delaney out*** The agent with the hide of a hippopotamus was the Max Markson, who remains amazingly and obdurately unashamed of taking advantage of such a disgraceful piece of behavior. (See "Corey Delaney Hired as Party Promoter," News.com, January 24, 2008, http://www.news.com.au/story/0,23599,23102828-2,00.html?from=mostpop.

100 ***this is the paradox that lies at the heart of groupthink*** D. J. T. Sumpter, "The Principles of Collective Animal Behaviour," *Philosophical Transactions of the Royal Society* 316 (2006): 7.

100 ***Janis thought [that the paradox could be overcome]*** Irving L. Janis, *Groupthink: Psychological Studies of Policy Decisions and Fiascos*, 2nd ed. (Boston: Houghton Miflin, 1982). Janis' argument is summarized in David J. T. Sumpter

and Stephen C. Pratt in "Quorum Responses and Consensus Decision Making," *Philosophical Transactions of the Royal Society B* 364 (2009): 744.

101 ***researchers Peter Gloor and Scott . . . specialize in the use of collective intelligence in business*** Peter Gloor and Scott Cooper, "The New Principles of a Swarm Business," *MIT Sloan Management Review* (Spring 2007): 81–84; Peter Gloor, *Swarm Creativity: Competitive Advantage Through Collaborative Innovation Networks* (New York: Oxford University Press, 2006); Peter Gloor and Scott Cooper, *Coolhunting—Chasing Down The Next Big Thing?* (NewYork: AMACOM, 2007).

102 ***Businesses [are taking] advantage of advances in complexity science*** Tom Lloyd discusses the spin-off company from the Santa Fe Institute that started it all ("When Swarm Intelligence Beats Brainpower," Santa Fe Institute, June 6, 2001, http://www.santafe.edu/~vince/press/swarm-intelligence.html). The company was bought by Nu-Tech Solutions in 2003 and continues to produce innovative ideas (www.nutechsolutions.com).

102 ***the worldwide cooperative movement*** www.cooperatives-uk.coop/live/images/cme_resources/Public/ots/Starting-a-Co-operative.pdf.

102 ***giant Migros supermarket chain*** Peter Gloor and Scott Cooper, "The New Principles of a Swarm Business," *MIT Sloan Management Review* (spring 2007): 81–84.

102 ***A stakeholder is defined as*** Peter Gloor and Scott Cooper, "The New Principles of a Swarm Business," *MIT Sloan Management Review* (spring 2007): 81–84.

103 ***how businesses like Amazon and eBay make their money*** Eric van Hest and Peter Vervest, "Smart Business Networks: How the Network Wins," *Communications of the Association for Computing Machinery* 50 (2007): 29–37.

The success of such businesses has been phenomenal compared to traditional businesses. eBay, for example, has a global customer base of 233 million at the time of writing and 88 million active users (buyers and sellers) per quarter ("Presentation on Q1 2009 Earnings Report of Ebay Inc.—Presentation Transcript, April 22, 2009, http://www.slide

share.net/earningreport/presentation-on-q1-2009-earning-report-of-ebay-inc#).

103 ***[Amazon and eBay] offer a business platform*** A. Kambil and E. van Heck, *Making Markets: How Firms Can Design and Profit from Online Auctions and Exchanges* (Boston: Harvard Business School Press, 2002).

103 ***provide the facilities for . . . a network*** Eric van Hest and Peter Vervest, "Smart Business Networks: How the Network Wins," *Communications of the Association for Computing Machinery* 50 (2007): 29–37.

104 ***IBM . . . spends $100 million per year*** H. W. Chesbrough, "Why Companies Should Have Open Business Models," *MIT Sloan Management Review* 48 (2007): 22–36.

104 ***Novartis . . . set up a venture fund*** Novartis, "Novartis Venture Fund Supports Innovative Start-Ups Despite Difficult Business Climate," media release, January 16, 2004, hugin.info/134323/R/930917/127663.pdf. See also Peter Gloor and Scott Cooper, "The New Principles of a Swarm Business," *MIT Sloan Management Review* (spring 2007): 83–84.

Chapter 7

I have drawn quite heavily on the work and writings of network science pioneers Duncan Watts and Albert-László Barabási in preparing this chapter. I acknowledge my debt, and in particular I thank Duncan for his helpful comments.

106 ***World Wide Web, and the Internet*** These two terms are often used synonymously, but they mean different things. The Internet is the linked network of computers, wires, software, satellites, etc. that carries messages and information via its hardware and software. The Web is a linked network of websites; it is one of the services carried by the Internet.

106 ***"a web without a spider"*** Albert-László Barabási, *Linked: How Everything Is Connected to Everything Else and What It Means for Business, Science, and Everyday Life* (New York: Plume, 2003), 219.

106 ***six degrees of separation*** This idea was popularized by the American playwright John Guare in a 1990 play and 1993 film of that name. Guare is said to have attributed his inspiration to Guglielmo Marconi's statement in his Nobel Prize acceptance speech that it would take 5.83 radio repeater stations to blanket the globe (Brian Hayes, "Graph Theory in Practice," *American Scientist* 88 [January–February 2000]: http://www.americanscientist.org/issues/pub/graph-theory-in-practice-part-i/5). If so, the inspiration was based on a false premise—Marconi said no such thing in his speech, which is available verbatim at http://nobelprize.org/nobel_prizes/physics/laureates/1909/marconi-lecture.pdf.

It was, in fact, the innovative Harvard and Yale psychologist Stanley Milgram (the same Milgram who arranged the staring-up-at-an-empty-window experiment described in chapters 3 and 7) who first developed the idea, although he never used the actual phrase (he called it "the small world problem" (*Psychology Today* 1 [1967]: 61–67). Credit must also go to sociologist Mark Granovetter, who argued that what holds a society together is not the strong ties within clusters but the weak ones between people who span two or more communities ("The Strength of Weak Ties," *American Journal of Sociology* 78 [1973]: 1,360–1,380).

The ultimate credit, though, may devolve to the Hungarian writer Frigyes Karinthy, whose 1929 short story *Láncszemek* (*Chains*) is based on the idea that everyone in the world is connected by at most five acquaintances. Barabási makes an interesting case that there may have been a chain between Karinthy and Milgram along which this idea was passed! (Albert-László Barabási, *Linked: How Everything Is Connected to Everything Else and What It Means for Business, Science, and Everyday Life* [New York: Plume, 2003], 37).

106 ***Samuel Johnson defined* network** Samuel Johnson, *Dictionary* (London: J. C. Robinson, 1828), 790. Like a number of Johnson's technical definitions, this one is rather mangled. A lady once asked him at a party why he had defined the pastern as the knee of a horse. His classic Johnsonian reply was "Ignorance, Madam. Pure ignorance" (quoted in James

Boswell, *The Life of Samuel Johnson*, Great Books of the Western World 44 [Chicago: Encyclopædia Britannica, 1952; reprinted in the 1990 edition as vol. 41]: 82).

107 ***Sociologists draw this set of connections as a sociogram*** Bernie Hogan, Juan Antonio Carrasco, and Barry Wellman, "Visualizing Personal Networks: Working with Participant-Aided Sociograms," *Field Methods* 19 (2007): 116–144.

108 ***nodes or links*** In the branch of mathematics known as "graph theory," these are respectively called *vertices* (plural of *vertex*) and *edges.*

108 ***A city . . . as a very complex network*** This example comes from John Holland (*Hidden Order: How Adaptation Builds Complexity* [New York: Basic Books, 1996]).

109 ***the point at which interconnectedness suddenly occurs*** Ray Solomonoff and Anatol Rapoport, "Connectivity of Random Nets," *Bulletin of Mathematical Biophysics* 13 (1951): 107–227. The discovery of this phenomenon is usually attributed to the famous mathematician Paul Erdös and his colleague Alfréd Rényi, who in fact discovered it independently some ten years later (see Albert-László Barabási, *Linked: How Everything Is Connected to Everything Else and What It Means for Business, Science, and Everyday Life* [New York: Plume, 2003], 245, for a list of references).

109 ***one random network is the highway system*** Albert-László Barabási, *Linked: How Everything Is Connected to Everything Else and What It Means for Business, Science, and Everyday Life* (New York: Plume, 2003), 71.

110 ***[Milgram's] best-known experiment*** Jeffrey Travers and Stanley Milgram, "An Experimental Study of the Small World Problem," *Sociometry* 32 (1969): 425–442.

110 **Six Degrees of Separation** John Guare, *Six Degrees of Separation: A Play* (New York: Vintage, 1990).

110 ***The simple statistics of a random network*** Milgram did not actually use this argument, although he quoted related arguments, in particular one used by Anotol Rapoport ("Spread of Information Through a Population with Socio-Structural

Bias," *Bulletin of Mathematical Biophysics* 15 [1953]: 523–543; A. Rapoport and W. J. Horvath, "A Study of a Large Sociogram," *Behavioral Science* 6 [1961]: 279–291). Readers of *Rock, Paper, Scissors* will recollect that Rapoport was also the author of the tit-for-tat–winning computer program in Robert Axelrod's famous evolution of cooperation competition.

A mathematically equivalent way of presenting the statistics is that used by Barabási: the number of links d in an average chain is proportional to the logarithm of the number of nodes N in the network. For the World Wide Web they found that $d = 0.35 + 2\log_{10}N$ (e.g., Albert-László Barabási, *Linked: How Everything Is Connected to Everything Else and What It Means for Business, Science, and Everyday Life* [New York: Plume, 2003], 33).

110 ***The Web . . . has nineteen degrees of separation*** R. Albert, H. Jeong, and A.-L. Barabási, *Nature* 401 (1999): 130–131.

111 ***some scholars have discovered serious flaws in Milgram's work*** Duncan Watts points out, for example, that unpublished research by Judith Kleinfeld, based on her survey of Milgram's original notes in the Yale archives, reveals that Milgram chose not to publish some data that did not support his hypothesis ("Small World Project Description," http://smallworld.columbia.edu/description.html).

111 ***Duncan Watts and Steve Strogatz . . . took a closer look*** "Collective Dynamics of 'Small-World' Networks," *Nature* 393 (1998): 440–442.

111 ***our social worlds could consist of tight, intimate networks*** Watts and Strogatz measured the tightness of the clusters using a cluster coefficient that measured the extent to which the people that a person knows also know each other. Say you have four friends. If they all knew each other, there would be six links between them. If some didn't know each other, there would be fewer links. If there were only four links, for example, then the cluster coefficient would be 4/6 = 0.67. ("Collective Dynamics of 'Small-World' Networks," *Nature* 393 [1998]: 440–442.)

111 ***just a few random links . . . created the same effect*** It's like passing a message from hand to hand through a

crowd, in contrast to calling someone on your cell phone who you know is standing next to the person that you want to get the message to. Another analogy is a wave at a football game, which relies on responding to the actions of one's near neighbors. Just two distant people in the crowd with cell phones, though, could coordinate things to produce two waves, traveling in opposite directions. I have never seen it done, but it offers all sorts of interesting possibilities for interference between the waves, like interference between the waves produced by two passing ships.

112 ***Friendship networks*** See, for example, the Friend Wheel that is available on Facebook and other social networking sites, such as http://Thomas-fletcher.com/friendwheel/.

113 ***Hollywood actor Kevin Bacon*** Watts and Strogatz used Bacon because there is an Internet game, "Six Degrees of Kevin Bacon" (see http://oracleofbacon.org), that has extensive data about Bacon's connections to other actors. The impetus for creating the game was a television interview in which Bacon claimed to have worked with practically everyone in Hollywood, or someone who has worked with them.

In an interview regarding the television program *Six Degrees*, Bacon revealed that he was not happy about his name being used in this way at first (see www.SixDegrees.org). He has turned it to advantage, however, by setting up the website SixDegrees.org, which builds on the popularity of the small world phenomenon to create a charitable social network and inspire people to give to charities online.

113 ***the neural network of the nematode worm* Caenorhabditis elegans** This little worm, barely a millimeter long, was selected for extensive study by Nobel Prize–winning developmental biologist Sydney Brenner in 1974. Brenner's plan (now carried through) was to learn more about how organisms develop by tracing the lineage of every cell in the worm's body. This would be impossible in humans, because we all have different numbers of cells, and also we have a high turnover of some cells. Every nematode has exactly the same number of cells (1,031 in the adult male, including 302 neurons), and it was possible to map the way they

are connected because it remains the same, generation after generation.

113 ***Duncan Watts advised me of recent work [that has] distinguished between two versions of the small world hypothesis*** Sharad Goel, Roby Muhammad, and Duncan Watts, "Social Search in 'Small-World' Experiments" (paper presented at the 18th International World Wide Web Conference (WWW2009), April, 20–24, 2009, Madrid, Spain).

113 ***there is a high probability that there will be a short chain of connection*** The length of the chain is approximated by the logarithm of the population size.

114 ***Watts and his coworkers adapted Milgram's experiment*** Peter Sheridan Dodds, Roby Muhamad, and Duncan J. Watts, "An Experimental Study of Search in Global Networks," *Science* 301 (2003): 827–829.

114 ***example of a chain letter*** Daniel W. VanArsdale, "Chain Letter Evolution," January 7, 2007, http://www.silcom.com/~barnowl/chain-letter/evolution.html. For the letter itself, see http://www.silcom.com/~barnowl/chain-letter/archive/me1935-05-20br_sd_club.htm.

116 ***modern equivalents include chain e-mails [and] text messages*** See, for example, a press release from the Bahrain telecommunications company Batelco concerning a hoax text message where customers were offered free air time if they passed the message to ten friends ("Batelco Customer Announcement: SPAM SMS Alert," May 20, 2008, http://www.arabianbusiness.com/press_releases/detail/19258.

116 ***Chain letters offer . . . substantial rewards*** One incentive with a sting in its tail is described by Robert Louis Stephenson in his story *The Bottle Imp*. The imp of the story has the power to grant wishes, but the owner of the bottle must sell it *at a loss* to someone else before they die, or their soul will burn in Hell. The story starts when the price, which started at millions of dollars, has dropped to just eight dollars. How it ends, you will have to find out for yourself.

117 ***closest to the target in terms of lattice distance*** Jon M.

Kleinberg, "Navigation in a Small World," *Nature* 406 (2000): 845.

117 ***knowledge of network ties and social identity*** Duncan J. Watts, Peter Sheridan Dodds, and M. E. J. Newman, "Identity and Search in Social Networks," *Science* 296 (2002): 1,302–1,305. The clever mathematical approach in this paper is really worth a look and should be relatively easy to follow for anyone with an understanding of high school algebra.

118 ***Murphy's Law of Management*** This and similar 80:20 rules (some practical, and some, like Murphy's Law itself, meant to make a point in a humorous fashion) can be traced back to the Italian economist Vilfredo Pareto, who observed that 80 percent of the land in Italy was owned by 20 percent of the population. Business management thinker Joseph M. Huran generalized the principle in 1937 to "the vital few and the trivial many" with such examples as "80% of the problem is caused by 20% of the causes" (John F. Reh, "Pareto's Principle: The 80-20 Rule—Complete Information," About.com, http://management.about.com/cs/generalmanagement/a/Pareto081202.htm). Huran called it the Pareto Principle in honor of its Italian originator.

118 ***the "champagne glass" effect*** Xabier Gorostiaga, "World Has Become a 'Champagne Glass'—Globalization Will Fill It Fuller for Wealthy Few," *National Catholic Reporter* January 27, 1995, http://findarticles.com/p/articles/mi_m1141/is_n13_v31/ai_16531823/.

118 ***The champagne glass shape of the graph denotes a power law*** Mathematicians refer to this as a "scale free" law. "Scale free" means that the shape of a graph remains the same when the vertical and horizontal axes are contracted or expanded in the same proportion. To give an example, when Pareto (see above) discovered his 80:20 rule, he found that it applied, not only to the whole of Italy, but to any particular smaller-scale area as well.

118 ***When Barabási . . . looked at the distribution of links*** Barabási and Albert thought that the U.S. electricity grid also followed a power law, although subsequent analysis showed that they were mistaken in this case (L. A. N. Amaral et al.,

Proceedings of the National Academy of Sciences of the U.S.A. 97 [2000]: 11,149–11,152). As well as the networks cited by Barabási, there are many other examples of power law distributions, including the distribution of species size in nature, the sizes of businesses and cities, and the sizes of natural events like forest fires and earthquakes (Duncan J. Watts, "The 'New' Science of Networks," *Annual Reviews of Sociology* 30 [2004]: 243–270).

120 ***similar power laws*** Duncan J. Watts, "The 'New' Science of Networks," *Annual Reviews of Sociology* 30 (2004): 243–270.

120 ***the Matthew effect*** Albert-László Barabási and Réka Albert, "Emergence of Scaling in Random Networks," *Science* 286 (1999): 509–512. The term was coined by the well-known sociologist Robert K. Merton to describe the way in which credit is often given to the senior, better-known scientist in a collaboration, even when it was the juniors who produced the ideas ("The Matthew Effect in Science," *Science* 159 [1968]: 56–63).

Merton was a great coiner of phrases; also to his credit are *self-fulfilling prophecy, role model,* and *unintended consequences.*

120 ***"a new actor . . . is more likely to be cast in a supporting role"*** Albert-László Barabási, *Linked: How Everything Is Connected to Everything Else and What It Means for Business, Science, and Everyday Life* (New York: Plume, 2003), 511.

121 ***analysis revealed that the Matthew effect would produce a power law distribution*** This distribution is "the inevitable consequence of self-organization due to local decisions made by individual [nodes], based on information that is biased toward the more visible (richer) [nodes], irrespective of the nature and origin of this visibility" (Albert-László Barabási, *Linked: How Everything Is Connected to Everything Else and What It Means for Business, Science, and Everyday Life* [New York: Plume, 2003]).

The emergence of hubs in networks is paralleled in physics by the emergence of order from chaos in the form of a phase transition, such as that which occurs when a relatively disordered liquid freezes to form a highly ordered crystalline solid.

121 ***power law distribution of connections in a growing network*** The various possibilities, and the underlying meaning (if any) of the widespread occurrence of power law networks has been discussed in detail by Evelyn Fox Keller ("Revisiting 'Scale-free' Networks," *Bioessays* 27 [2005]: 1,060–1,068).

121 ***these possibilities have been incorporated into the burgeoning crop of network models*** See Duncan J. Watts, "The 'New' Science of Networks," *Annual Reviews of Sociology* 30 (2004): 252, for a good list of early examples.

121 ***preferential attachment is driven by the product of a node's fitness and connectivity*** G. Bianconi and A.-L. Barabási, "Competition and Multiscaling in Evolving Networks," *Europhysics Letters* 54 (2001): 436–442.

121 ***one node can grab* all *of the links*** G. Bianconi and A.-L. Barabási, "Bose-Einstein Condensation in Complex Networks," *Physical Review Letters* 86 (2001): 5,632–5,635.

121 ***Barabási advanced Microsoft's Windows operating system as an example*** Albert-László Barabási, *Linked: How Everything Is Connected to Everything Else and What It Means for Business, Science, and Everyday Life* (New York: Plume, 2003), 104.

122 ***Another example . . . is that of military censorship*** Richard Feynman experienced this when he was working on the Manhattan Project. The story of how he and his wife Arlene managed to undermine the system by various ruses is told in Richard P. Feynman, *Surely You're Joking, Mr. Feynman! Adventures of a Curious Character* (New York: Norton, 1985).

123 ***"bow tie" theory [of directed networks]*** A. Broder et al., "Graph Structure in the Web," http://www9.org/w9cdrom/160/160.html (2000).

123 ***power law networks . . . are very stable*** Random networks fall apart after a critical number of nodes have been removed so that the average connectivity is less than one per node. For power law networks there is no such critical

threshold if the "power" is less than three, which is the case for most real-life networks (Reuven Cohen et al., "Resilience of the Internet to Random Breakdowns," *Physical Review Letters* 85 [2000]: 4,626; D. S. Callaway et al., "Network Robustness and Fragility: Percolation on Random Graphs," *Physical Review Letters* 85 [2000]: 5,468–5,471).

123 ***the loss of one keystone species can trigger . . . collapse*** References in Albert-László Barabási, *Linked: How Everything Is Connected to Everything Else and What It Means for Business, Science, and Everyday Life* (New York: Plume, 2003), 262.

123 ***These strictures apply to any network made up of one-way links*** This was proved by Sergey Dorogovtsev, José Mendes, and A. N. Samukhin, quoted in Albert-László Barabási, *Linked: How Everything Is Connected to Everything Else and What It Means for Business, Science, and Everyday Life* (New York: Plume, 2003), 169.

124 ***a radical rethinking about how viral diseases . . . spread*** The description that I give here is necessarily simplified, and the actual calculations can be both highly complicated and sensitive to some of the assumptions made. For a comprehensive update, see Lisa Sattenspiel, *The Geographic Spread of Infectious Diseases: Models and Applications*, Princeton Series in Theoretical and Computational Biology (Princeton: Princeton University Press, 2009).

124 ***the story of how the Black Death made it to the English village of Eyam*** John Clifford, *Eyam Plague 1665–1666*, self-published 1995; revised 2003, available through the Eyam Museum, http://www.eyammuseum.demon.co.uk/booklist.htm.

125 ***AIDS hubs*** French Canadian flight attendant Gaëtan Dugas, known in the history of the AIDS epidemic as "patient zero," was claimed to be the first of these, with somewhere between 2,500 and 20,000 sexual partners from the time of diagnosis in 1982 until his death in 1984 (Randy Shilts, *And the Band Played On* [New York: St. Martin's Press, 2000]). Recent research has demonstrated, however, that AIDS arrived in America via a single infected immigrant from Haiti in about 1969 (M. Thomas P. Gilbert et al., "The Emergence

of HIV/AIDS in the Americas and Beyond," *Proceedings of the National Academy of Sciences of the U.S.A.* 104 [2007]: 18,566–18,570; see also "AIDS Virus Invaded U.S. from Haiti: Study," Reuters, October 30, 2007, http://www.reuters.com/article/scienceNews/idUSN2954500820071030).

125 ***Mathematical analyses*** The indispensable pioneering work in this field is Martin A. Nowak and Robert M. May, *Virus Dynamics: Mathematical Principles of Immunology and Virology* (Oxford: Oxford University Press, 2009).

125 ***concentrate on the hubs*** Romualdo Pastor-Satorras and Allesandro Vespignani, "Immunization of Complex Networks," *Physical Review E* 65 (2002): 1–8, http://www-fen.upc.es/~romu/Papers/immuno.pdf.

125 ***swine flu . . . whole school was closed down*** Valerie Elliott, "Alleyn's School Is Fifth Closed by Swine Flu as UK Cases Rise to 27," *The Times*, May 5, 2009, http://www.timesonline.co.uk/tol/news/uk/article6222747.ece.

128 ***the role of opinion leaders in the formation of public opinion*** Elihu Katz and Paul Lazarsfeld, *Personal Influence: The Part Played by People in the Flow of Mass Communications* (Glencoe, IL; The Free Press, 1955).

128 ***Marketers in particular have adopted this theory*** Christophe Van den Bulte and Joshi V. Yogesh, "New Product Diffusion with Influentials and Imitators," *Marketing Science* 26 (2007): 400–421 (see http://knowledge.wharton.upenn.edu/papers/1322.pdf).

128 ***Recent work in network science . . . has largely debunked it*** Duncan J. Watts and Peter Sheridan Dodds, "Influentials, Networks and Public Opinion Formation," *Journal of Consumer Research* 34 (2007): 441–458; Duncan Watts, "Challenging the Influentials Hypothesis," *Measuring Word of Mouth* 3 (2007): 201–211.

128 ***if we want to kick off a cascade of influence*** For a summary of the many factors that can influence the initiations and propagation of cascades, see Duncan J. Watts, "A Simple Model of Global Cascades on Random Networks," *Proceedings of the National Academy of Sciences of the U.S.A.* 99 (2002): 5,766–5,771.

129 ***the history of viral marketing*** Duncan J. Watts, Jonah Peretti, and Michael Frumin, "Viral Marketing for the Real World," Collective Dynamics Group, 2007, cdg.columbia.edu/uploads/papers/watts2007_viralMarketing.pdf.

130 ***"contagious media" contest*** Duncan J. Watts, Jonah Peretti, and Michael Frumin, "Viral Marketing for the Real World," Collective Dynamics Group, 2007, cdg.columbia.edu/uploads/papers/watts2007_viralMarketing.pdf.

130 ***[Watts and his collaborators] investigated [the chances for success of a song] in the competitive music download market*** Matthew J. Salganik, Peter Sheridan Dodds, and Duncan J. Watts "Experimental Study of Inequality and Unpredictability in an Artificial Cultural Market," *Science* 311 (2006): 854–856.

Chapter 8

I have drawn considerably on the work of Gerd Gigerenzer, Director of the Center for Adaptive Behavior and Cognition at the Max Planck Institute for Human Development in Berlin, and that of the international ABC (Center for Adaptive Behavior and Cognition) Research Group of which he is a member. They have pioneered the idea that "less is more" and other simple heuristics can actually be superior to more complex ways of thought and have shown us why.

133 ***[The history of gold prospecting in] Sofala*** Joyce Pearce, *Gold Nuggets Galore at Sofala from 1851* (Bathurst: Western Printers, 1982).

134 ***classical decision theory*** This is an enormous subject that is divided into normative ("a theory about how decisions *should* be made in order to be rational") and descriptive, which is about how decisions *are* actually made. Its tentacles penetrate into economics (purchasing and investment decisions), politics (voting and collective decision making), psychology (*how* we make decisions), and philosophy (what are the requirements for rationality in decisions?). I touch on some of these topics in this book but make no attempts to

cover the whole field. There are many standard textbooks, but for the interested reader who would like something in plain and understandable language I highly recommend the summary by Sven Ove Hansson, "Decision Theory: A Brief Introduction," 1994, revised 2005, http://www.infra.kth.se/~soh/decisiontheory.pdf.

135 ***One of my favorite cartoons*** The cartoon is by James Thurber (one of my favorite cartoonists) *New Yorker*, December 29, 1934, http://www.cartoonbank.com/product_details.asp?sid=39517.

135 ***Benjamin Franklin . . . "moral algebra"*** Letter to Joseph Priestley, September 19, 1772, see http://homepage3.nifty.com/hiway/dm/franklin.htm.

136 ***Gigerenzer and his colleague Daniel Goldstein asked students*** Daniel G. Goldstein and Gerd Gigerenzer, "Models of Ecological Rationality: The Recognition Heuristic," *Psychological Review* 109 (2002): 75–90, http://www.dangoldstein.com/papers/RecognitionPsychReview.pdf. See also Gerd Gigerenzer, Peter M. Todd, and the ABC Research Group, *Simple Heuristics That Make Us Smart* (Oxford: Oxford University Press, 1999), 43.

Gigerenzer has recently repeated the experiment using "Detroit or Milwaukee" instead of "San Diego or San Antonio," because in the last few years the population of San Antonio has grown to match that of San Diego (Gerd Gigerenzer, *Gut Feelings: The Intelligence of the Unconscious* [New York: Viking, 2007]).

136 ***a more apt title for our species would be* Homo heuristicus** Gerd Gigerenzer and Henry Brighton, "Homo Heuristicus: Why Biased Minds Make Better Inferences," *Topics in Cognitive Science* 1 (2009): 107–143.

Classical economic theory begins with the assumptions that we are not just *Homo sapiens* ("thinking man") but also *Homo omnisciens* ("all-knowing man") and *Homo omnipotens* (able to process complex information in a short space of time). Together these add up to the impossible dream of *Homo economicus*, the totally rational being who inhabits economics textbooks, who is always aware of all the factors involved in a decision, and who is always able to come to the best and most self-serving conclusion. Economists are only just starting to come to grips with the fact that *Homo economicus*, like bigfoot and the yeti, has never yet been seen in the real world. *Homo heuristicus*, though, is everywhere.

136 ***force field analysis*** was pioneered by the German psychologist Kurt Lewin in the late 1940s (K. Lewin, *Field Theory in Social Science* [New York: Harper and Row, 1951]).

137 ***a force field diagram*** See, for example, http://www.valuebasedmanagement.net/methods_lewin_force_field_analysis.html.

137 ***Gigerenzer and his colleagues have identified a total of ten heuristics*** Gerd Gigerenzer "Why Heuristics Work," *Perspectives on Psychological Science* 3 (2008): 20–29. The headings that I give in the text are very slightly rephrased and rearranged from the list in this seminal paper.

138 ***Five Additional Heuristics***

For completeness, I list here the additional heuristics that Gigerenzer gives.

1. Spread Your Bets Evenly

Instead of choosing one alternative over another, allocate your resources equally to each.

When investing in a portfolio of stocks and shares, this simple strategy consistently outperforms all efforts to balance the investments according to the expected returns in terms of the Sharpe ratio (a measure of the excess return, or risk premium, per unit of risk in an investment asset or a trading strategy, certainty-equivalent return, or turnover). This is basically because "the gain from optimal diversification is more than offset by estimation error" (Victor DeMiguel, Lorenzo Garlappi, and Raman Uppal, "1/N" (paper presented at the 2006 European Financial Association Meeting, Zurich, 2006).

2. Take the Do-Nothing Default

If there is a default option to do nothing, then do nothing.

When it comes to organ donation, for example, it has been suggested that requiring people who *do not* want to donate their organs to carry a card (so that donation becomes the default option) could save thousands of lives a year (E. J. Johnson and D. G. Goldstein, "Do Defaults Save Lives?" *Science* 302 [2003]: 1,338–1,339).

3. Tit-for-Tat

In a situation where there is a choice between cooperation and noncooperation and the situation is likely to come up in the future, cooperate on the first encounter. In subsequent encounters, do whatever the other party did during the first encounter. If they cooperated, keep cooperating. If they didn't cooperate, stop cooperating.

As readers of *Rock, Paper, Scissors* will know, this simple strategy won a computer competition to produce cooperation in an ongoing prisoner's dilemma situation (Robert Axelrod, *The Evolution of Cooperation* [New York: Basic Books, 2006]). It has since been improved upon by the win-stay, lose-shift strategy—that is, stick with your strategy (cooperate or don't cooperate) if it is serving you well; shift to the opposite strategy on the next round if your present strategy fails you in this round (Martin A. Nowak and Karl Sigmund, "A Strategy of Win-Stay, Lose-Shift That Outperforms Tit-for-Tat in the Prisoner's Dilemma Game," *Nature* 364 [1993]: 56–58). Win-stay, lose-shift avoids the endless cycles of retaliation that tit-for-tat can lead to.

4. Imitate the Majority

Take the course of action that the majority of your peer group is taking.

See R. Boyd and P. J. Richerson, *The Origin and Evolution of Cultures* (New York: Oxford University Press, 2005).

No need for much comment here, other than to point out that this can lead equally to consensus via a quorum response or to the deadly trap of groupthink (see chapter 6).

5. Imitate the Successful

Follow the example of those who have succeeded.

Again, see Boyd and Richerson.

This strategy is likely to succeed only if you have the same qualities that the successful person you are imitating has. Usually this includes a capacity for hard work, which immediately eliminates many teenagers who want to become celebrities.

138 ***Gigerenzer and Goldstein [used the "less-is-more" recognition principle] to guide investments in stocks and shares*** Gerd Gigerenzer, Peter M. Todd, and the ABC Research Group, *Simple Heuristics That Make Us Smart* (Oxford: Oxford University Press, 1999), 61–72.

138 ***the German stock exchange*** Deutsche Börse.

139 ***A group of Canadian sociologists [made] the name Sebastian Weisdorf famous overnight*** Larry L. Jacoby et al., "Becoming Famous Overnight: Limits on the Ability to Avoid Unconscious Influences of the Past," *Journal of Personality and Social Psychology* 56 (1989): 326–338.

140 ***American Apparel . . . was sued by Woody Allen*** The suit was settled out of court (Ed Pilkington, "Woody Allen Reaches $5m Settlement with Head of American Apparel," *The Guardian*, May 18, 2009, http://www.guardian.co.uk/world/2009/may/18/woody-allen-american-apparel-settlement).

140 ***[the hat] had a tantalizing air of familiarity*** It was, in fact, the one that she had worn at our wedding.

140 ***The advantages of forgetting were first recognized by . . . William James*** William James, *The Principles of Psychology*, vol. 1 (New York: Holt, 1990), 679–680.

141 ***By forgetting the less desirable options, we are better able to use recognition*** Some psychologists argue that forgetting prevents out-of-date information from interfering with the recall of currently useful information (E. L. Bjork and R. A. Bjork, "On the Adaptive Aspects of Retrieval Failure in Autobiographical Memory," in *Practical Aspects of Memory*, vol. 2, edited by M. M. Gruneberg, P. E. Morris, and R. N. Sykes [London: Wiley, 1988]: 283–288). Others argue that forgetting previous goals makes it easier to focus on current goals (E. M. Altmann and W. D. Gray, "Forgetting to Remember," *Psychological Science* 13 [2002]: 27–33).

141 ***forgetting [helps us] to use recognition to pick out the best option*** Lael J. Schooler and Ralph Hertwig, "How For-

getting Aids Heuristic Inference," *Psychological Review* 112 (2002): 610–628.

Forgetting could also have helped me as a scientist. I would have been better off if I had been able to follow the example of the late Enrico Fermi, father of the atomic pile, and forget most of the equations that I had been taught. Like most of my fellow students, I spent painful hours memorizing them for exams, and even more painful hours dredging them from my memory in the course of my professional work. Fermi had a different approach. He worked out that most problems in physics fall into one of seven categories, and memorized the equations relevant to these categories. With that, he could forget the rest of the equations, and rely on recognizing the category into which any particular problem would fall.

142 **Blink** Malcolm Gladwell, *Blink: The Power of Thinking Without Thinking* (New York: Little, Brown, 2005).

Sue Halpern makes the point that Gladwell's "fevered" book has a "subject and intent [that] closely follow Gigerenzer's [earlier book *Simple Heuristics That Make Us Smart*]" (*New York Review of Books* 52, April 28, 2005). Critics have argued that some of Gladwell's illustrations are flawed (see, for example, Howard Davies' review in *The Times*, February 5, 2005, http://entertainment.timesonline.co.uk/tol/arts_and_entertainment/books/article510197.ece.

Gladwell applied the "instinct" idea to a much wider range of situations than Gigerenzer and his colleagues thought appropriate. Professor Gigerenzer has pointed out in personal correspondence that the question that "Gladwell does not ask is how can we know when an intuition will be good or bad? To answer this question one needs first to ask what is the underlying process—such as heuristics—a question he also does not ask. Gigerenzer and his colleagues have analyzed in some details the situations where a heuristic works and where it is not successful—the study of what they call the *ecological rationality* of a given heuristic."

142 ***The underlying science of the fluency decision strategy*** Lael J. Schooler and Ralph Hertwig, "How Forgetting

Aids Heuristic Inference," *Psychological Review* 112 (2002): 610–628.

142 ***"Quick decisions are unsafe decisions"*** Often attributed to the Greek playwright Sophocles, even though nothing like it appears in his collected works. Its first appearance was in a collection of around 700 maxims written by a freed Roman slave called Publilius Syrus around 40 b.c.e. Other maxims included "A rolling stone gathers no moss" and "There are some remedies worse than the disease" (John Bartlett, *Familiar Quotations*, 13th ed. [Boston: Little, Brown, 1955], 44–45).

Other versions of the maxim include "Nothing can be done at once hastily and prudently" (Bartlett, *Familiar Quotations*) and "Measure twice and cut once" (a favorite maxim of my current workshop supervisor).

143 ***Charles Darwin [listed a series of] pros and cons [for and against marriage]*** He did this in a notebook that is now on display at his house, known as Down House, in Kent, England. The full text of his list is available in *The Complete Works of Charles Darwin Online*, November 2008, http://darwin-online.org.uk/content/frameset?viewtype=text&itemID=CUL-DAR210.8.2&pageseq=1, together with an image of the original document.

143 ***The arguments for marriage occupied considerably more space*** This is my interpretation, reached after rearranging Darwin's lists according to a pro and con template.

143 ***decision theorist Robyn Dawes demonstrated [that tallying] could sometimes outperform much more complicated approaches*** R. M. Dawes and B. Corrigan, "Linear Models in Decision Making," *Psychological Bulletin* 81 (1974): 95–106; R. M. Dawes, "The Robust Beauty of Improper Linear Models in Decision Making," *American Psychologist* 34 (1979): 571–582.

143 ***Dawes['] . . . conclusions were greeted by the scientific community with howls of outrage*** Gerd Gigerenzer and Henry Brighton, "Homo Heuristicus: Why Biased Minds Make Better Inferences," *Topics in Cognitive Science* 1 (2009): 111.

144 ***When researchers tried it across a range of twenty . . . categories*** J. Czerlinski, G. Gigerenzer, and D. G. Gold-

stein, "How Good Are Simple Heuristics?" in Gerd Gigerenzer, Peter M. Todd, and the ABC Research Group, *Simple Heuristics That Make Us Smart* (Oxford: Oxford University Press, 1999), 97–118.

145 ***levelwise Tallying*** The basic idea of Levelwise Tallying goes back to a conjecture by Paul Meehls that "an unweighted sum of a small number of 'big' variables will, on the average, be better than regression equations [i.e., elaborate statistical analysis LF]" (Gerd Gigerenzer, Peter M. Todd, and the ABC Research Group, *Simple Heuristics That Make Us Smart* [Oxford: Oxford University Press, 1999], 112). The phrase itself was introduced by Jean-François Bonnefon et al., in "Qualitative Heuristics for Balancing the Pros and Cons," where the authors tested a total of eight different "Tallying" heuristics, and concluded that Levelwise Tallying produced the best results (*Theory and Decision* 65 [2008]: 71–95).

145 ***Studies have shown that tallying succeeds best*** H. J. Einhorn and R. M. Hogarth, "Unit Weighting Schemes for Decision Making," *Organizational Behavior and Human Performance* 13 (1975): 171–192.

145 ***evolution [dictates that female guppies] prefer males that are the brightest orange*** L. A. Dugatkin and J. J. Godin, "How Females Choose Their Mates," *Scientific American* 278 (1998): 46–51, quoted in Gerd Gigerenzer, Peter M. Todd, and the ABC Research Group, *Simple Heuristics That Make Us Smart* (Oxford: Oxford University Press, 1999), 81–82.

146 ***we are attracted to people whose facial appearance and body shape are symmetrical*** Gillian Rhodes et al., "Facial Symmetry and the Perception of Beauty," *Psychonomic Bulletin and Review* 5 (1998): 659–669; William M. Brown et al., "Fluctuating Asymmetry and Preferences for Sex-Typical Bodily Characteristics," *Proceedings of the National Academy of Sciences of the U.S.A.* 105 (2008): 12,938–12,943.

146 ***"take-the-best" is expected to do well*** Robin M. Hogarth and Natalia Karelaia, 'Take-the-Best' and Other Simple Strategies: Why and When They Work 'Well' with Binary Cues," *Theory and Decision* 61 (2006): 205–249; H.

Brighton, "Robust Inference with Simple Cognitive Models," in *Between a Rock and a Hard Place: Cognitive Science Principles Meet AI-Hard Problems*, edited by C. Lebiere and B. Wray, Association for the Advancement of Artificial Intelligence Technical Report No. SS-06-03, 17–22 (Menlo Park: AAAI Press, 2006).

146 ***Satisficing*** This term was coined by the economist Herbert Simon in his essay "A Behavioral Model of Rational Choice," *Quarterly Journal of Economics* 69 (1955): 99–118.

147 ***we accept a reasonable level of satisfaction*** Barry Schwartz et al., "Maximizing Versus Satisficing: Happiness Is a Matter of Choice," *Journal of Personality and Social Psychology* 83 (2002): 1,178–1,197; Michael Byron, ed., *Satisficing and Maximizing; Moral Theorists on Practical Reason* (Cambridge: Cambridge University Press, 2004).

147 ***the secretary problem*** John P. Gilbert and Frederick Mosteller, "Recognizing the Maximum of a Sequence," *Journal of the American Statistical Association* 61 (1966): 35–73; Thomas Dudey and Peter Todd, "Making Good Decisions with Minimal Information: Simultaneous and Sequential Choice," *Journal of Bioeconomics* 3 (2001): 195–215.

The history of the secretary problem and its many ramifications for choice in everyday life is given in a beautiful article by Thomas S. Ferguson, *Statistical Science* 4 (1989): 282–289. The history goes back at least as far as astronomer Johannes Kepler's search for a new wife in 1611.

148 ***This 37 percent rule*** is actually a "$1/e$" rule, where e is the transcendental number 2.718281828. . . . A full mathematical description of the general problem is given by Dietmar Pfeiffer in "Extremal Processes, Secretary Problems and the 1/e Law," *Journal of Applied Probability* 27 (1989): 722–733.

148 ***heuristics experts Peter Todd and Geoffrey Miller*** P. M. Todd and G. F. Miller, "From Pride and Prejudice to Persuasion: Satisficing in Mate Search," in Gerd Gigerenzer, Peter M. Todd, and the ABC Research Group, *Simple Heuristics That Make Us Smart* (Oxford: Oxford University Press, 1999), 287–308.

149 ***Could we use the same strategy in choosing a life partner?*** P. M. Todd and G. F. Miller, "From Pride and Prejudice to Persuasion: Satisficing in Mate Search," in Gerd Gigerenzer, Peter M. Todd, and the ABC Research Group, *Simple Heuristics That Make Us Smart* (Oxford: Oxford University Press, 1999), 287–308 (a highly recommended book and chapter).

149 ***They may meet our aspirations—but do we meet theirs?*** This sort of balance arises in a lot of situations in life. One prime example concerns the logic used by many British railway networks when they claim that most of their trains are on time. Their "satisficing" definition of "on time" is "within ten minutes," which may keep them happy, but does not satisfice too many commuters.

149 ***they may not be as happy with you as you are with them*** P. M. Todd and G. F. Miller, "From Pride and Prejudice to Persuasion: Satisficing in Mate Search," in Gerd Gigerenzer, Peter M. Todd, and the ABC Research Group, *Simple Heuristics That Make Us Smart* (Oxford: Oxford University Press, 1999), 300–308.

150 ***A simple example [of daily temperature variation] will serve to illustrate the difference*** This example is taken from Gerd Gigerenzer and Henry Brighton, "Homo Heuristicus: Why Biased Minds Make Better Inferences," *Topics in Cognitive Science* 1 (2009): 107–143, where it is discussed in more detail.

150 ***the "bias-variance" dilemma*** S. Geman, E. Bienenstock, and R. Doursat, "Neural Networks and the Bias-Variance Dilemma," *Neural Computation* 4 (1992): 1–58; Gerd Gigerenzer and Henry Brighton, "Homo Heuristicus: Why Biased Minds Make Better Inferences," *Topics in Cognitive Science* 1 (2009): 107–143.

151 ***Simple rules that allow them to respond*** Kathleen M. Eisenhardt and Donald N. Sull, "Strategy as Simple Rules," *Harvard Business Review* (January 2001): 106–116. I have relied heavily on this article in preparing this section.

151 ***Eisenhardt and Sull [studied dozens of companies]*** Including Enron (!), whose collapse was *not* due to the use of simple rules.

Chapter 9

155 ***episode of* Peanuts** This dialog originally appeared in the print form of the strip, and was subsequently used in the 1969 film *A Boy Named Charlie Brown*, directed by Bill Melendez, written by Charles M. Schulz, and produced by Cinema Center Films and Lee Mendelson Films for National General Pictures.

156 ***imagination can take many forms [both in science and the arts]*** See, for example, Arthur Koestler, *The Act of Creation* (New York: MacMillan, 1964) and the response by Nobel Laureate Sir Peter Medawar, "The Act of Creation," in *The Strange Case of the Spotted Mice and Other Classic Essays on Science* (Oxford: Oxford University Press, 1996: 40–51).

156 ***large areas of the brain . . . are used for the perception of visual and aural patterns*** Giorgio Ganis, William L. Thompson, and Stephen M. Kosslyn, "Brain Areas Underlying Visual Mental Imagery and Visual Perception: An fMRI Study," *Cognitive Brain Research* 20 (2004): 226–241; Mark Jung-Beeman, "Bilateral Brain Processes for Comprehending Natural Language," *Trends in Cognitive Sciences* 9 (2005): 512–518.

157 ***Mendeleev . . . wanted to find some organizing principle*** Dmitri Ivanovich Mendeleev, *Mendeleev on the Periodic Law: Selected Writings, 1869–1905*, edited by William B. Jensen (Mineola, NY: Dover, 2005); Professor Mendeleev, "The Periodic Law of the Chemical Elements," Faraday lecture, Chemical Society in the Theatre of the Royal Institution, June 4, 1889, *Journal of the Chemical Society* 55 (1889): 634–656, http://web.lemoyne.edu/~giunta/mendel.html.

157 ***discovery of the periodic table*** There is a nice image of one of the early versions at http://en.wikipedia.org/wiki/File:Mendeleev_Table_5th_II.jpg.

158 ***a search for patterns and regularities in the depths of complexity*** The philosopher David Hume argued that the "laws of nature" that scientists so proudly claim to have uncovered are no more than regularities, with no deep underlying meaning (*A Treatise of Human Nature* [1739], edited

by L. A. Selby-Bigge [London: Oxford University Press (1888), reprinted 1960]). Others have argued that Hume's views were just the opposite of this, and that careful reading reveals that he was a "necessitarian" (Tom Beauchamp and Alexander Rosenberg, *Hume and the Problem of Causation* [New York: Oxford University Press, 1981]). The whole issue is explored by David Armstrong in *What Is a Law of Nature?* (Cambridge: Cambridge University Press, 1983).

158 ***Murray Gell-Man [found that the different] components of the atomic nucleus . . . fit a pattern of symmetry*** The pattern of symmetry is called the eight-fold way (Michael Riordan, *The Hunting of the Quark: A True Story of Modern Physics* [New York: Simon & Schuster, 1987]) by loose analogy with the Buddhist teaching of the Noble Eightfold Path (e.g., Bhikkhu Bhodi, *The Noble Eightfold Path: The Way to the End of Suffering* (Kandy, Sri Lanka: Buddhist Publication Society, 2006).

158 ***Jim Watson . . . used cut-out cardboard shapes*** James D. Watson *The Double Helix: A Personal Account of the Discovery of the Structure of DNA*, edited by Gunther S. Stent, Norton Critical Editions (New York: Norton, 1980). I particularly recommend this edition because it contains copies of most of the reviews published at the time of the first edition, including some in which the reviewers have quite spectacularly shot themselves in the foot.

158 "***nobody ever got anywhere by seeking out messes***" James D. Watson *The Double Helix: A Personal Account of the Discovery of the Structure of DNA*, edited by Gunther S. Stent, Norton Critical Editions (New York: Norton, 1980)

158 ***Mathematical biologist James Murray certainly succeeded*** John M. Gottman et al., *The Mathematics of Marriage: Dynamic Nonlinear Models* (Cambridge: MIT Press, 2002). See also James D. Murray, "The Marriage Equation: A Practical Theory for Predicting Divorce and Scientifically-Based Marital Therapy" (public lecture, Institute for Mathematics and its Applications, University of Minnesota, November 18, 2004, http://www.ima.umn.edu/public-lecture/2004-05/murray/; J. D. Murray, "Modelling the Dynamics of Marital Interaction: Divorce Prediction and

Marriage Repair," in *Mathematical Biology I: An Introduction*, 3rd ed. (New York: Springer, 2002), 146–174; James Murray, "Mathematics in the Real World: From Brain Tumours to Saving Marriages," (Bakerian Prize lecture, Royal Society, London, March 26, 2009, http://royalsociety.org/page.asp?id=8303).

159 ***data miners do something similar*** See for example, the very readable Y. Peng et al., "A Descriptive Framework for the Field of Data Mining and Knowledge Discovery," *International Journal of Information Technology and Decision Making* 7 (2008): 639–682.

Those interested in more detail could try Jiawei Han and Micheline Kamber's *Data Mining: Concepts and Techniques*—a massive, 770-page tome that contains a real wealth of detail for professional data miners on how to drink from the firehose of data (a phrase coined by M. Mitchell Waldrop in *Complexity* [London: Penguin Books, 1994], 63), albeit with a worrying paucity of pages about the protection of private data (2nd ed. [San Francisco: Morgan Kaufmann, 2006]).

160 ***[Simon Newcomb] noticed that the book of logarithms . . . had very dirty pages at the front*** Simon Newcomb, "Note on the Frequency of Use of the Different Digits in Natural Numbers," *American Journal of Mathematics* 4 (1881): 39–40.

Newcomb was also professor of mathematics and astronomer at the United States Naval Observatory, Washington D.C., at the time—a remarkable achievement for someone with practically no conventional schooling, although somewhat aided by the fact that many staff with Confederate sympathies had left at the start of the Civil War (Bill Carter and Merri Sue Carter, *Simon Newcomb: America's Unofficial Astronomer Royal* [Matanzas, Cuba: Matanzas Publishing, 2006]).

160 ***Newcombe discovered that there was a mathematical rationale*** It needs no more than high school mathematics to understand the reasoning, although the argument is rather too long to reproduce here. See Jon Walthoe, Robert Hunt, and Mike Pearson, "Looking Out for Number One," *Plus*

(September 1999): http://plus.maths.org/issue9/features/benford/, for good examples of the law in action.

161 ***Frank Benford found that the law applied*** Frank Benford "The Law of Anomalous Numbers," *Proceedings of the American Philosophical Society* 78 (1938): 551–572.

161 ***Benford's law could be used as a test for financial fraud*** Mark Nigrini, "I've Got Your Number: How a Mathematical Phenomenon Can Help CPAs Uncover Fraud and Other Irregularities," *Journal of Accountancy* 187 (May 1999): 15–27. For subsequent developments, see Nigrini's website, www.nigrini.com.

161 ***President Bill Clinton's published tax accounts*** Unpublished data of Mark Nigrini (personal communication, May 2009).

161 ***Many other applications of Benford's law are now being explored*** Economist Hal Varian (chief economist at Google at the time of writing) was first off the mark in 1972 with his slightly tongue-in-cheek suggestion that Benford's Law could be used to check the reasonableness of the forecasting models used in planning applications ("Benford's Law," *The American Statistician* 26 [1972]: 65). For other applications in economics, see Stefan Günnel and Karl-Heinz Tödter, "Does Benford's Law Hold in Economic Research and Forecasting?" *Deutsches Bundesbank Discussion Paper Series 1: Economic Studies*, No. 32 (2007). For applications in the pharmaceutical industry, see Fabio Gambarara and Oliver Nagy, "Benford Distribution in Science," Eidgenössische Technische Hochschule, Zurich, July 17, 2004, http://www.socio.ethz.ch/education/mtu/downloads/gambarara_nagy_2004_mtu.pdf.

161 ***The ancient Babylonians and Assyrians believed that there were relationships between events in the sky and events here on Earth*** Erica Reiner, "Babylonian Celestial Divination," in Noel M. Swerdlow, *Ancient Astronomy and Celestial Divination* (Cambridge: MIT Press, 2000), 22–23; Francesca Rochberg, "Heaven and Earth: Divine-Human Relations in Mesopotamian Celestial Divination," in *Magic in History: Prayer, Magic and the Stars in the Ancient and Late Antique World*, edited by Scott Noegel, Joel

Walker, and Brannon Wheeler (University Park: Pennsylvania State University Press, 2003), 180.

161 ***the post hoc fallacy*** A occurs before B, therefore A is the cause of B. For a brief but more detailed description and examples, see http://www.nizkor.org/features/fallacies/post-hoc.html.

162 ***a book of omens*** There were eventually seventy such books Erica Reiner, "Babylonian Celestial Divination," in Noel M. Swerdlow, *Ancient Astronomy and Celestial Divination* (Cambridge: MIT Press, 2000), 22.

162 ***Frank Plumpton Ramsey was a Cambridge mathematician*** with a quite incredible intellect. He once learned enough German in a week (!) to produce an English translation of Ludwig Wittgenstein's *Tractatus* that is still regarded as a model of its kind. His translations were sometimes very much to the point. One of Wittgenstein's often-quoted conclusions is "Whereof one cannot speak, one must be silent." Ramsey translated this as "what you can't say, you can't say, and you can't whistle it either." F. P. Ramsey, "General Propositions and Causality," in *F. P. Ramsey: The Foundations of Mathematics*, edited by R. B. Braithwaite (London: Routledge, 1931).

David Mellor, emeritus professor of philosophy at the University of Cambridge, has been particularly active in bringing Ramsey's life and work to public attention. See D. H. Mellor, "Cambridge Philosophers I: F. P. Ramsey," *Philosophy* 70 (1995): 243–262, and the radio program "Better Than the Stars," http://sms.cam.ac.uk/media/20145.

162 ***the proportion of its income that a nation should prudently save*** F. P. Ramsey, "A Mathematical Theory of Saving," *Economic Journal* 38 (1928): 543–559.

162 ***John Maynard Keynes called it "one of the most remarkable contributions"*** John Maynard Keynes, "Frank Plumpton Ramsey," in *Essays in Biography* (New York: Harcourt, Brace, 1933).

163 ***Ramsey's theorem*** popped up as an almost irrelevant side issue when Ramsey was working out his economic theories. It is technically described in terms of a graph, which means in mathematical terms a set of points (vertices) connected

by lines (edges). The theorem says:

Let *Kn* be a complete graph on *n* vertices (that is, each point is connected by lines to all the other points). Then, for integers *m*, $n \geq 2$, there exists a least positive integer $R = R(m.n)$ such that every edge-coloring of *KR*, using the colors red and blue, must admit either a *Km* in which every edge is red, or a *Kn* such in which every edge is blue (adapted from Tim Le Saulnier "On the Asymptotic Behavior of Ascending Waves" (citeseerx.ist.psu.edu)).

One of the more surprising consequences of Ramsey's theorem is that, if we take a set of whole numbers and list them in any order, there will always be a certain number that form an ascending or descending series (this is known as Van der Waerden's theorem). If we have 101 whole numbers, for example, no matter how we order them there will still be at least 11 somewhere in the set that form such a series.

163 ***"The rate of saving multiplied by the marginal utility of money"*** F. P. Ramsey, "A Mathematical Theory of Saving," *Economic Journal* 38 (1928): 543–559.

164 ***graph theory*** This is described in Alexander Soifer, *The Mathematical Coloring Book: Mathematics of Coloring and the Colorful Life of Its Creators* (New York: Springer, 2009). The stories of the people concerned are indeed fascinating, but be warned—the mathematics is not always at the simple level implied by the title.

164 ***Ramsey numbers*** The Ramsey number $R(m,n)$ gives the solution to the party problem, which asks the minimum number of guests $R(m,n)$ that must be invited so that at least *m* will know each other or at least *n* will not know each other (Eric W. Weisstein, "Ramsey Number, *MathWorld*—A Wolfram Web Resource, http://mathworld.wolfram.com/RamseyNumber.html)

Stanislaw Radziszowski maintains an up-to-date list of progress in calculating Ramsey numbers at http://www.combinatorics.org/Surveys/#DS1.

INDEX